Physiology 1
Laboratory Text

Human Physiology

Santa Rosa Junior College
Biological Sciences Department

Manual Prepared by:
Susan Wilson, PhD, and Nick Anast, MA

Published by Arbor Crest Publishing 

July 2022–11

Dedicated to SRJC Anatomy and Physiology Instructor Nick Anast

"Beyond all that sort of technical information, what is equally as important is love, compassion, empathy. Can you project that into the world? Is that something we do?
Could you feel love from across the room from somebody? I think we can.
I think we can project our love across the country.
I think we could project our love across the world if it's strong enough.
So, start practicing; practice a way of love and harmonizing."

Nick Anast, 1959 – 2015

Table of Contents

Introduction to the Lab

The laboratory exercises and lecture material are designed to compliment each other. Direct laboratory experience leads to a clearer understanding of the concepts and facts presented in the text and lectures. Both are an integral part of the course. There is a great deal of material to cover in this course, and lectures move quickly. Laboratory time is meant to proceed more slowly. It is a time for hands on experience with physiological concepts, and a time to learn techniques that may prove useful in your future careers. It is also a time to make friends with classmates and form study partnerships.

The directions for the lab exercises in this manual are intentionally brief and to the point. These directions will be supplemented by verbal directions given by the lab instructor before you begin each exercise. You may also read more extensive *directions* in manuals kept on the physiology reference shelf in the lab, and obtain more information on *concepts* from your text.

Your job each day is to come to lab prepared, which means that you have read over the experiment in your lab manual. If directions are not clear, please ask questions and check reference books. Your work in the lab will be graded in two ways. There is a report to complete after every lab and turn in to be graded. Please note that you may share raw data, and discuss results and problems, but you may not share the writing of a lab report. Copying someone else's report is considered cheating. There will also be a lab exam portion of each midterm exam, based on the following. You are responsible for knowing all of the information in the lab manual. You should be able to recognize (identify, know the function of) every reagent, piece of equipment, slide, model that you use. Questions can also be based on the lab reports and on data collected. With this in mind, you should take notes in your lab manual as you go through the experiments, so that you have something to review from.

There are several labs in which data from the whole class are pooled and you must analyze these results. After data are collected and hand written in class data worksheets, data will be entered into a classroom computer for calculation and subsequent distribution to the whole class.

Physiology Lab Rules and Procedures Agreement

After reviewing this document with your instructor, you will sign a document stating that you agree to all the following. This will be kept on file in the biology department at SRJC.

1. Be properly prepared to do the experiment. Read the written procedures *in advance* and understand what you are going to do.

2. We use corrosive chemicals for some of the experiments we conduct during lab. On these days wear personal protective clothing such as long pants and closed toe shoes.

3. Perform the experiments as directed. Do not do anything which is not part of an approved experimental procedure. Follow all instructions given by the instructor including but not limited to handling of chemicals including their disposal, cleaning up after chemical use, and proper disposal of sharp instruments called "sharps". Never pipet by mouth.

4. Know where to find the SRJC Emergency Preparedness Handbook and become familiar with it. In the case of any emergency from your cell phone call the Campus Police at 707-527-1000 (from campus phones dial ext. 1000).

5. Learn the locations and operation of emergency equipment. This includes sinks, fire alarm, fire extinguisher, eyewash station, and first aid supplies.
6. Always act in a responsible manner. No horseplay or fooling around is allowed.
7. Report all accidents, injuries, and close calls to the instructor immediately.
8. Give you instructor a list of any allergies and/or medical conditions. If the experiment involves something to which you are allergic consult with the instructor.
9. If you have any health issue of concern or that might pose a risk to others please let the instructor know. Anything you tell the instructor will be kept in strict confidence.
10. Never take supplies, equipment, or books out of the laboratory.
11. **Food and Beverage in Lab**. No eating is allowed while in lab. Drinking is generally allowed while in lab but only in containers with screw-on lids. There are, however, some exceptions. They include, during the ***Renal Lab*** and the ***Blood Lab***, due to using bodily fluids, no food or beverages are allowed in the lab room. **NOTE: At any time, new regulations may be put in place as Santa Rosa Junior College deems necessary for the protection of everyone.**
12. Clean up all spills immediately. This includes water.
13. At the end of *each* lab class the benches are to be thoroughly cleaned and all equipment is cleaned and put away as instructed.

Laboratory Exercise 1

Scientific Data & Method

Introduction

Science is a dynamic process, a way of knowing about our world. This process is called the **scientific method**. The first step in this way of knowing is **observation**, which is followed by questions, formation of a **hypothesis** that suggests a possible explanation of the observation or an answer to the question, an **experiment** to test the hypothesis, and a **conclusion** about the validity of the hypothesis. As you learn new facts during this physiology course, remember the process used to learn these facts, and remember that all scientific "facts" are subject to revision as more is learned.

During many of the laboratory exercises conducted this semester you will be collecting and analyzing data. This is an essential part of physiology, and differs from an anatomy lab where one simply makes observations and learns the names of specific structures. There are several skills necessary for handling **data**, which will be reviewed in this laboratory exercise. First, some very basic ideas: the word data is plural (substitute the word results in your mind); every data point or number written down in physiology must have a unit associated with it (with a few exceptions, such as specific gravity). Thus if you write on a laboratory report or lab practical exam "10", you will be marked wrong. Is this 10 miles? 10 kilograms? 10 beats/minute? 10 chocolate bars? The number 10 is meaningless without units.

Scientists use the metric system, scientific notation, ratios and proportions, and statistical analysis to present and manipulate data they have collected. You must become familiar with these mathematical expressions and relationships in order to correctly present the data you will collect during this semester. This lab should be a review, as these concepts are part of most high school curricula.

Objectives

- Review the steps of the scientific method.
- Become familiar with the metric system: learn prefixes but do not memorize conversion factors.
- Make standard conversions.
- Write numbers in scientific notation.
- Solve simple physiological word problems using ratios.
- Distinguish continuous versus discrete physiological data.

Materials

Conversion tables
Calculators

Procedures

Conversions: Scientists use the **metric system.** You must become familiar with this system and be able to make conversions between the "English system" and the metric system of measurement. The metric system utilizes units that are based on the decimal system and are related to one another by some power of ten. The term denoting a metric unit of measurement usually contains a prefix indicating the power of ten. Table 1.1 indicates prefixes used in the metric system. Tables 1.2, 1.3, 1.4 indicate the units of linear measure, weight and volume in both systems. Conversions between the two systems can be made using proportions, as described on the following page. To convert between Celsius and Fahrenheit scales the following formulas are used: $C = 0.56\,(F - 32)$

$F = (1.8 \times C) + 32$

Scientific Notation: Scientists also frequently deal with very large and very small numbers when an observation must be described quantitatively. Such numbers are cumbersome to manipulate if written in the form of a decimal. **Scientific notation** simplifies the expression and manipulation of such numbers. Scientific notation is a floating–point system of numerical expression in which numbers are expressed as products consisting of a number between 1 and 10 multiplied by an appropriate power of 10. To convert any number to scientific notation write the number such that there is one number (between 1 and 10) to the left of the decimal point, and the rest of the numbers (other than zeros) to the right. The power of ten indicates where the decimal point is correctly located. When any number greater than 10 is expressed by scientific notation, the decimal point is moved to the left until the number has a value between 1 and 10, and that number is by a positive power of 10 equal to the number of places the decimal point was moved to the left. For example: $10{,}263 = 1.0263 \times 10^4$. When any number less than 1 is expressed in scientific notation, the decimal point is moved to the right until the number has a value between 1 and 10, and that number is multiplied by a negative power of 10 equal to the number of places the decimal was moved to the right. For example: $0.01467 = 1.467 \times 10^{-2}$.

Ratios and Proportions: Expression and computation of scientific data is often most easily accomplished by using **ratios** and **proportions**. A ratio is an expression that compares two numbers or quantities by division. Whenever two quantities are expressed as a ratio, they must have the same units. One cannot compare the weights of two animals if the first is expressed in pounds and the second in kilograms. A proportion is a mathematical statement of the equality of two ratios. A proportion can be stated as follows:

$$\frac{A}{B} = \frac{C}{D}$$

This formula says that A is to B as C is to D. Mathematically it says that A x D = B x C. If three of the quantities are known, the fourth can be calculated. For example if a muscle contraction is mechanically recorded on graph paper moving at a speed of 50mm/second and the trace for the contraction covers 5mm, one can determine the time duration of the muscle contraction.

$$\frac{50\text{mm}}{1\text{ sec}} = \frac{5mm}{x\text{ sec}} \qquad \text{rearranging gives} \quad 50x = 5 \quad \text{and} \quad x = 0.1\text{ sec}$$

Averages: Comparison of groups of scientific data often includes calculation of the arithmetic mean or **average**. The formula for the mean is $x = \sum x / N$

Statistics and Data: Scientists often want to then determine if there is a significant difference between two means. This involves statistical analysis, which will not be covered in this course. When collecting data from humans, which is what we will be doing for most of the laboratory exercises, there is often a large degree of variability, compounded by the fact that the data will be collected by many observers, with a wide range of skills. Thus almost none of our data will show statistically significant differences, and we will be noting only trends or suggestions of differences.

Data are often plotted as a **graph**. There are two kinds of graphs frequently used, line graphs and bar graphs. Data that are continuous can be plotted as a line graph, whereas discontinuous data are plotted as a bar graph. Look through your text for samples of both kinds of graphs.

Practice these mathematical manipulations by doing the problems outlined in your laboratory report.

Table 1.1 Metric Prefixes				
Prefix	Abbreviation	Meaning	Factor	Decimal
tera	T	one trillion	10^{12}	1,000,000,000,000
giga	G	one billion	10^{9}	1,000,000,000
mega	M	one million	10^{6}	1,000,000
myiz	my	ten thousand	10^{4}	10,000
kilo	k	one thousand	10^{3}	1,000
necto	h	one hundred	10^{2}	100
deka	da	ten	10^{1}	10
uni		one	10^{0}	1.0
deci	d	one–tenth	10^{-1}	0.1
centi	c	one–hundredth	10^{-2}	0.01
milli	m	one–thousandth	10^{-3}	0.001
micro	μ	one–millionth	10^{-6}	0.000001
nano	n	one–billionth	10^{-9}	0.000000001
pico	p	one–trillionth	10^{-12}	0.000000000001
femto	l	one–quadrillionth	10^{-15}	0.000000000000001

Table 1.2 Metric units of linear measure		
Metric Unit	Definition	English equivalent
megameter (Mm)	10^{6} meters	621.37 Miles
myriameter (mym)	10^{4} meters	6.2137 Miles
kilometer (km)	10^{3} meters	0.62137 miles
hectometer (hm)	10^{2} meters	328.0833 feet
dekameter (dam)	10 meters	32.80833 feet
meter (m)	basic unit of reference	39.37 inches
decimeter (dm)	10^{-1} meter	3.937 inches
centimeter (cm)	10^{-2} meter	0.3937 inch
millimeter (mm)	10^{-3} meter	0.03937 inch
micrometer (μ)	10^{-6} meter	0.00003937 inch
nanometer (nm)	10^{-9} meter	0.00000003937 inch
angstrom (Å)	10^{-10} meter	0.000000003937 inch
picometer (pm)	10^{-12} meter	0.00000000003937 inch
femtometer (fm)	10^{-15} meter	0.00000000000003937 inch

Table 1.3 Metric units of weight		
Metric unit	Definition	English equivalent (avoirdupois)
metric ton	10^6 grams	2204.62 pounds
kilogram (kg)	10^3 grams	2.20462 pounds
hectogram (hg)	10^2 grams	0.220462 pound
dekagram (dag)	10 grams	0.35274 ounce
gram (g)	basic unit of reference	0.035274 ounce
decigram (dg)	10^{-1} gram	0.0035274 ounce
centigram (cg)	10^{-2} gram	0.00035274 ounce
milligram (mg)	10^{-3} gram	0.000035274 ounce
microgram (µg)	10^{-6} gram	0.000000035274 ounce
nanogram (ng)	10^{-9} gram	0.000000000035274 ounce
picogram (pg)	10^{-12} gram	0.000000000000035274 ounce

1.4 Metric units of volume		
Metric unit	Definition	English equivalent (U.S.)
myraliter (myl)	10^4 liters	2641.7 gallons
kiloliter (kl)	10^3 liters	264.17 gallons
hectoliter (hl)	10^2 liters	26.417 gallons
dekaliter (dal)	10 liters	10.567 quarts
liter (l)	basic unit reference	1.0567 quarts
deciliter (dl)	10^{-1} liter	0.10567 quart
centiliter (cl)	10^{-2} liter	0.010567 quart
milliliter (ml)	10^{-3} liter	0.0010567 quart
microliter (µl)	10^{-6} liter	0.0000010567quart
nanoliter (nl)	10^{-9} liter	0.0000000010567quart
picoliter (pl)	10^{-12} liter	0.0000000000010567 quart
femtoliter (fl)	10^{-15} liter	0.0000000000000010567 quart

Human Physiology Lab 1 Name ____________________

Scientific Data

1. Use a small metric rule to measure the length of the line shown below. Record the measurement in millimeters and centimeters.

______________________________ __________ mm

__________ cm

2. Record your weight in pounds, your height in inches; convert these measurements to kilograms and centimeters (dream values ok to use!):

weight: _________ lbs = _________ kg height: _________ inches = _________ cm

3. Compute the following conversions:

242 mg = __________ g 3450 ml = __________ L

6 kg = __________ g 243 mm = __________ cm

4 kg = __________ lbs 10 C = __________ °F

0.83 cm = __________ mm 72 F = __________ °C

4. Solve the following proportions for x:

$6/36 = x/48$ $x =$ __________ 9:72 as x:64 $x =$ __________

$24/144 = 18/x$ $x =$ __________ $x/27 = 17/81$ $x =$ __________

5. Express the following numbers using scientific notation:

1563 = __________ 0.364 = __________ 0.0042 = __________ 54.63 = __________

6. At rest, the left ventricle of the heart pumps 5.0 liters of blood per minute. Blood flow to the kidneys is approximately 1,200 ml per minute at rest. Assuming a proportionate increase in renal blood flow, what will be the blood flow to the kidneys if the heart pumps 7.0 liters/min? *Show your work!*

7. An electrocardiogram is recorded on mm grid paper moving at a speed of 25mm/sec. If the distance between beats as recorded is 20 mm, what is the subject's heart rate in beats per minute? *Show your work!*

Laboratory Exercise 2

Homeostasis

Introduction

The regulatory mechanisms of the body help to maintain a state of dynamic constancy of the internal environment known as **homeostasis**. The body maintains homeostasis through negative feedback mechanisms. Homeostasis is constantly threatened by changes in the external environment and even by activities of the body itself, such as strenuous exercise. Maintaining homeostasis requires a *sensor*, which detects a change in one of the many parameters which are monitored. This information is sent to an *integrator*, in which a set point for each parameter has been established. The integrator in turn activates an *effector* (muscle or gland), which induces changes opposite to those that activated the sensor. Activation of the effector thus compensates for the initial disturbance, so there is only a slight deviation from the state of constancy. Since a disturbance in homeostasis initiates events that lead to changes in the opposite direction, this mechanism is called a **negative feedback loop**.

Objectives

- Define homeostasis and explain how negative feedback loops maintain homeostasis.
- Observe normal fluctuations in one homeostatically regulated parameter.
- Observe the effects of a functioning negative feedback loop.

Materials

Your body!
Clock or watch

Procedures

1. Gently press your index and middle fingers (not your thumb) against the radial artery in your wrist until you feel a pulse.
2. Count the number of pulses in a 15 second interval, record your data. Multiply these values by 4 to obtain the pulse rate/minute
3. Take your pulse for 15 seconds, pause for 15 seconds, then take your pulse for another 15 second interval. Repeat this procedure over a 5 minute period, obtain 10 data points.
4. Enter your data in the lab report table and plot it as a line graph on the grid provided in the report.
5. Record your average pulse rate in your lab report and on the class data chart that will be put on the board. Notice the variation in the data collected.

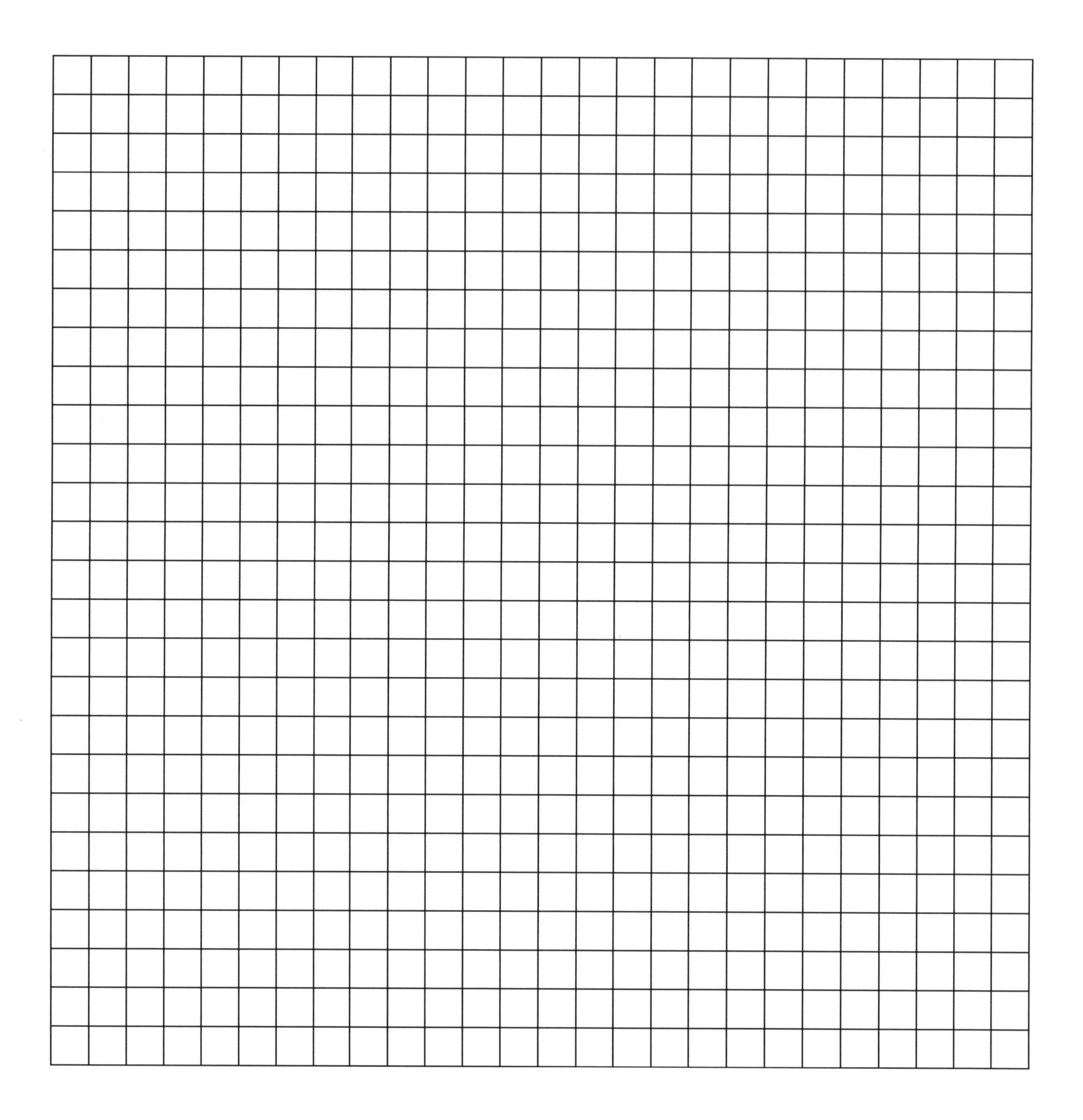

Human Physiology Lab 2 Name ______________________________

Homeostasis and Negative Feedback

1. Data

Measurement	1	2	3	4	5	6	7	8	9	10
Beats per 15 seconds										
Beats per minute										

Calculate your average pulse rate: ____________ beats per minute (record here and on board).

2. Graph your data on the graph provided.

3. Define the term homeostasis and describe the components of negative feedback mechanisms that operate to maintain homeostasis.

4. Explain how your graph of pulse rate measurements illustrates the presence of a negative feedback control mechanism and dynamic constancy.

5. Describe two observations that you could make based on the class data.

6. What factors might influence the amount of variation observed in the class?

Laboratory Exercise 3

Cells, Tissues, Microscope

Introduction

This lab is a review of information that students should know from the physiology pre–requisite course, introductory biology. Some instructors may omit this lab.

The cell is the basic unit of structure and function in the body. It is essential in physiology to understand that everything that is done in the body or by the body is done by cells. The major parts of the cell include a plasma membrane, nucleus, and cytoplasm. The cytoplasm includes organelles, most of which are membrane bound, and a liquid matrix filled with dissolved solutes called cytosol. The plasma membrane, nucleus and cytoplasm can be identified in cells using the light microscope. Most organelles can only be visualized with an electron microscope. We will use cell models to view the organelles.

Cells with similar specializations of structure and function are grouped together to form tissues. In turn, tissues are grouped together to form larger units of structure and function called organs. In this lab we will review the four major tissue types: epithelial, connective, muscle, nervous. Many organs are made from all four tissue types. We will review one such organ, the skin.

In addition to learning the basic parts of a cell, the names and functions of the principal organelles, and the four tissue types, this laboratory also provides an opportunity to review the parts of the microscope, and its proper care and use.

Objectives

- Review proper care of microscope.
- Review parts of microscope and their function.
- Differentiate cells, tissues, organs.
- Be able to identify the following cell organelles on model and give their functions:
  nucleus, nucleolus, ribosome, rough ER, smooth ER, Golgi complex, lysosome, mitochondria
- Review names and characteristics of the 4 major tissue types.
- Recognize 4 tissue types in one organ, the skin.

Materials

Microscope
Cell model

Procedures

Microscope: parts and care

When handling the microscope, treat it as you would a four day old Australian Shepherd puppy. USE EXTREME CARE! The microscope you are using is a compound microscope, which means that it has a combination of lenses. One lens (the ocular) is in the eyepiece and produces a tenfold (10X) increase in magnification. The other lens (the objective) is found closest to the object being viewed and will additionally magnify the object dependent on the specific power of that objective. The total magnification of the microscope is obtained by multiplying the magnifying power of the ocular times that of the objective lens. Be sure to learn the names and magnifying power of the 4 objectives (4X: low power, 10X: medium power, 40X: high power) as well as the names and functions of all the other microscope parts. Use Figure 3.1 for review.

The following rules for the care and use of microscopes are to be learned and practiced when using microscopes in this lab.

1. Carry the microscope with two hands, upright, close to your body. Do NOT drag it across lab bench!
2. Clean the lenses with lens paper before and after use – never with Kimwipes!
3. Always begin a microscope study using the low or scanning objective. Orient the visual area of interest into the center of the field of view. Have the slide close to the objective, and focus by moving the objective away from the slide with the course focus knob.
4. When using high power, only use the fine focus for adjustments.
5. Leave the lowest power objective in place when putting the microscope away.

Tissues

Look up in your text the major characteristics of each of the four major tissue types: epithelial, connective, muscle, nervous. Look at a drawing of skin in your text or a histology reference book, and be able to identify structures that are made of each of the four tissue types.

Cell Model

Be able to recognize the structure and explain the function of the following cell organelles: nucleus, nucleolus, rough and smooth endoplasmic reticulum, mitochondria, ribosomes, Golgi apparatus, lysosomes. Use the cell model key and your textbook as guides.

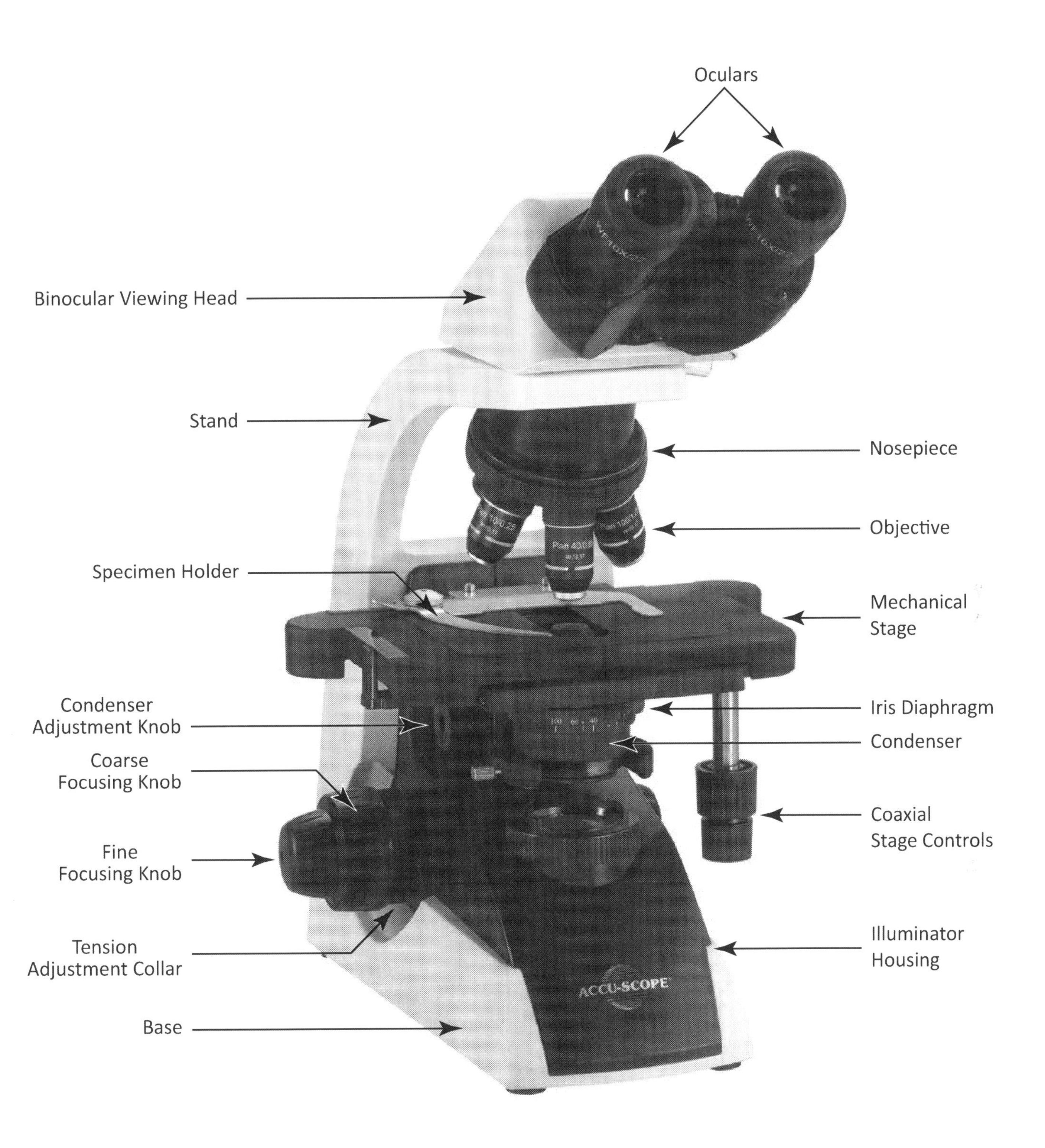

Figure 3.1–A compound microscope

Microscope Components and their Functions *		
Microscope Component	**Location or Description**	**Functions**
Ocular (eyepiece)	Uppermost series of lenses	Magnification
Nosepiece	Revolving assembly	Holds objectives
Scanning power objective	Shortest objective, magnifies 4x	Magnification
Other objectives	Microscope also has 10x, 40x & 63x objectives	Magnification
Condenser	Lens system located below central stage opening	Concentrates and directs light beam through specimen
Stage	Platform upon which specimens for examinations are placed	Specimen support
Mechanical stage	Control device that permits moving slides left to right, forward & backward on stage	Manipulation of specimen's location
Iris diaphragm	Located beneath stage in association with condenser unit, controlled with a lever	Regulates brightness or intensity of light passing through lenses
Coarse adjustment knob	Large knob on arm, below stage	Used for preliminary and coarse focusing by raising or lowering stage
Fine adjustment knob	Smaller knob, below stage	Used for final or fine focusing by raising or lowering stage
Condenser adjustment knob	Control knob located below stage	Used to obtain full illumination by raising or lowering condenser
Base	Heavy, bottom portion on which the instrument rests	Microscope support
In base illuminator	Light switch, rheostat	Turn on, adjust light intensity
Arm	Vertical, curved portion of microscope, used in carrying instrument	Microscope support

** This list is for reference purposes, the list of microscope parts to learn for exams is on the previous page*

Human Physiology Lab 3 Name ______________________________

Cells and Tissues

1. What is the function of the following parts of the microscope?

   stage ______________________________

   ocular ______________________________

   objectives ______________________________

   condenser ______________________________

   iris diaphragm ______________________________

   coarse adjustment knob ______________________________

   fine adjustment knob ______________________________

2. Give the total magnification when you use the high–dry power objective lens ___________

3. Define the term tissue:

   ____

   Define the term organ:

   ____

4. From the description provided, identify the organelle:

   major site of ATP production in cell ______________________________

   site of protein synthesis ______________________________

   contains digestive enzymes ______________________________

   location of genetic information ______________________________

   dense mass of protein & RNA in nucleus ______________________________

5. Many organs contain all four tissue types.

   Name the 4 tissue types; name a specific structure made of each tissue type and found in skin.

Tissue Type	Specific structure in skin
______________	______________
______________	______________
______________	______________
______________	______________

Laboratory Exercise 4

Enzyme Activity

Introduction

Catalysts are substances that accelerate chemical reactions without changing the nature of the reaction and without being altered by the reaction. **Enzymes** are biological catalysts and are made of protein. They function because they have a specific three–dimensional binding site to which the reactants or substrate molecules bind. This binding facilitates the formation or breaking of chemical bonds, i.e. chemical reactions. The shape of the binding or active site is determined by the amino acid sequence of the protein, and is different for each enzyme. Enzymes are therefore specific, interacting only with specific substrates and catalyzing only specific reactions.

Since enzymatic activity is dependent on the three-dimensional shape of polypeptide chains, changes of pH, temperature and salt concentration will affect the three-dimensional shape of enzymes and therefore their activity. Each enzyme has a pH optimum and a temperature optimum, and the enzyme activity diminishes considerably when conditions vary from the optima.

In order to study enzymes in the lab various factors must be considered. There must be a biological source of crude enzyme or a purified preparation; a supply of substrate; controlled conditions of pH, temperature, & concentration; and a way of measuring enzyme activity. In this lab the simplest conditions will be used. The reaction that we will study is the conversion of hydrogen peroxide to water and oxygen gas:

$$2H_2O_2 \rightarrow 2H_2O + O_2$$

This reaction can be catalyzed with an inorganic catalyst, iron, obtained from a rusty nail. A way to measure the progress of the reaction is to simply observe the production of oxygen bubbles in the solution. And the solution is merely the substrate, hydrogen peroxide, purchased from any drug store. Clinically and in the home, 3% hydrogen peroxide is used to disinfect wounds. This reaction will also occur slowly without a catalyst, which is why hydrogen peroxide is sold in brown bottles since light accelerates the reaction. The reaction can also be catalyzed with an enzyme called **catalase.** Catalase contains iron atoms bound in chemical structures called porphyrins, which are in turn bound to a protein, similar to the structure of hemoglobin. This reaction occurs 30,000 times faster with iron, and 100 million times faster with catalase. Catalase is found in many tissues within organelles called **peroxisomes** and is one of the most rapidly acting enzymes in the body. We will use a crude liver extract as our enzyme source, prepared by simply mincing fresh animal liver. The reaction rate will be observed by noting the production of oxygen bubbles. We can compare actions of inorganic and organic catalysts, and we expect to observe evidence for the difference in reaction rates noted above. We can also observe the effect of temperature on enzyme activity by comparing the rate of oxygen production with liver at room temperature and chilled liver.

Objectives

- Compare the effects of an inorganic and an organic catalyst.
- Become familiar with the role of enzymes in chemical reactions.
- Observe the effect of temperature on enzyme activity.
- Differentiate catalyst and enzyme; reactant and substrate.

Materials

Beakers
Rusty nails
Animal liver
Hydrogen peroxide

Procedures

1. Fill 2 small beakers up to 20 mls with hydrogen peroxide.
2. Immerse a rusty nail in one beaker, swirl gently, observe the effect of an inorganic catalyst.
3. Add minced liver to the second beaker and observe the effect of an organic catalyst, an enzyme.
4. A chilled third beaker, liver and hydrogen peroxide will be provided to compare the reaction rate with that observed at room temperature.

 Name ___________________________

Enzyme Activity

1. Describe what occurred when the rusty nail was placed in the beaker of hydrogen peroxide.

   Write the balanced equation that summarizes this reaction: ___________________________

2. How was the reaction different when chicken liver was placed in a beaker of hydrogen peroxide?

   What was the effect of cold on the rate of the catalase reaction?

3. Differentiate the terms catalyst and enzyme; reactant and substrate:

4. Name 3 factors that could cause an enzyme to denature.

   ____________________ ____________________ ____________________

5. Where is catalase found in cells? ___________________________

   What is the relation, if any, between rusty nails and catalase?

Laboratory Exercise 5

Osmosis

Introduction

Throughout the body there are several types of membranes, which serve as barriers between different body or fluid compartments. They serve to confine certain substances and processes to specific locations. Some of these membranes are cell membranes, i.e. the phospholipid bilayer, including the plasma membrane which surrounds each cell and the membranes that surround organelles. Other membranes are composed of cells or tissues. Some examples of these are mucous membranes, serous membranes, the respiratory membrane in lungs, the filtration membrane in kidneys, and the capillary walls (endothelium). All of these membranes not only separate compartments but must allow some substances to pass through while restricting the movement of others. The transport of various chemicals across these membranes is of critical importance for the maintenance of homeostasis. The transport of substances across membranes involves both *physical processes* (filtration, diffusion, osmosis) as well as *biochemical processes* (carrier-mediated transport). In this lab we will study some of the physical processes associated with transport of substances across membranes.

Some definitions are important before the lab work can be described. A substance that enters into solution or becomes dissolved in a liquid is called a **solute**. The liquid in which it dissolves is called a **solvent**. **Filtration** is the removal of particles suspended in a solution by passing the solution through a selectively permeable membrane. Simple **diffusion** is the continual random movement of particles in a liquid or gas, and *net* diffusion occurs when there is measurable movement of particles from an area of greater concentration to an area of lesser concentration. This will occur until there is a uniform distribution of the diffusing substance. The rate of diffusion is directly proportional to the concentration gradient and the temperature, but inversely proportional to the size (molecular weight) of the solute and the distance traveled. **Osmosis** is the net diffusion of water across a selectively permeable membrane.

Next, note some classic examples of where each of these processes occur in the body. Blood is **filtered** in the kidneys to form urine. Simple diffusion occurs across plasma membranes if there is a concentration difference across the membrane, and if the membrane is permeable to the solute in question. **Osmosis** can occur across all cell membranes, with potentially devastating results. Cells and multicellular organisms have evolved several mechanisms to prevent the damaging effects of osmosis: cell shrinking or swelling. Cell shrinking is called **crenation**, and cell swelling can lead to **lysis**, i.e. the bursting of the plasma membrane.

Osmosis is difficult to understand because it is not intuitive. The solvent (water) is moving across a membrane, not the solute. Imagine a solution divided into two compartments by a membrane. If the membrane is completely permeable to both solute and solvent, both will diffuse across freely, and there will be no concentration difference between the two sides of the membrane. Suppose, however, that the membrane is permeable only to solvent but not to solute. Solvent (water) will diffuse from the region where there is less solute / more water to the side with more solute / less water. This process is called osmosis. To help keep this straight one can think of water moving down *its* concentration gradient, from high to low. Or, to put it more simply, *solutes suck!*

This description of a semipermeable membrane separating solutions of unequal solute nature and concentration is a description of cells bound by plasma membranes. Cells are filled with solutes that cannot cross the membrane, they are osmotically active, they "suck in" water. In the course of evolution, cells have had to evolve mechanisms for dealing with this problem. Plant cells build cell walls, and animal cells pump NaCl across their membranes and live in homeostatically regulated interstitial fluid. In lab you will observe how cell walls protect plant cells, and how damaging are the effects of placing animal cells in solutions that are not "osmotically matched to their cytoplasm". You must also become familiar with the correct terms for such solutions. If the concentration of osmotically active solute is the same on both sides of a membrane, the solutions are said to be **isotonic**. If another solution has less osmotically active solute, it is **hypotonic**. If it has more, it is **hypertonic.** The suffix for these three terms is derived from the word tonicity, which refers to the osmotic pressure of a solution relative to that of blood. Such a solution is **300 mosm/L**. When an animal cell is placed in a hypertonic solution it shrivels (becomes **crenated**); when placed in a hypotonic solution it swells and bursts (**lyses**).

Note that when a substance such as salt dissolves, it dissociates into two ions, each of which is osmotically active. This does not happen with substances like sugars. Because each dissociated part of a dissolved solute is osmotically active, the osmolarity of a solution must consider the number of solute particles formed when ionic compounds dissolve.

Objectives

- Understand the concepts of osmosis and tonicity.
- Observe effects of placing *Elodea* leaves in isotonic, hypotonic, hypertonic solutions.
- Recognize the shape and significance of crenated and lysed red blood cells.

Materials

Dialysis tubing
Sucrose and saline solutions
Digital scale
Elodea leaves (or red onion)

Procedures

Part 1. Osmosis with artificial membranes

1. You will use dialysis tubing, which is an artificial, semipermable membrane.
2. Make little bags using the tubing and add 25 ml of the solution provided by lab instructor.
3. Weigh the bag (time 0 reading) and immerse in a second solution contained in your beaker.
4. Every 15 minutes remove the bag, gently blot it dry, and weigh to the nearest tenth of a gram, for a total of 4 readings.
5. Determine whether your bag has lost or gained weight.

After all the data are collected, you will be asked to determine which solution, pure water, 15% sucrose or 30% sucrose, was in your bag and which was in your beaker.

Part 2. Osmosis in living cells

Some cells can be directly observed as they respond to changes in salt and water in their immediate environment. Observing red blood cells is relevant to human physiology. If these cells are placed in a hypotonic solution they may swell and lyse. When placed in a hypertonic solution, they may lose water and become shrunken and wrinkled; these cells are said to be crenated. Red blood cells would be ideal to observe but often present practical problems for use in lab. Plant cells, however, can be readily observed, providing a more assured opportunity to observe osmosis in living cells. Plant cells behave differently than animal cells in response to osmotic stress. They do not lyse in hypotonic solutions because of the rigidity of their cell wall. However, in hypertonic solutions the cell will shrink away from the more rigid cell wall. This is called plasmolysis, not crenation. *Elodea* leaves will be used to observe osmosis in plant cells. *Elodea* is a freshwater plant with numerous, relatively small leaves which are only two cells thick. This allows the cytoplasm of individual cells to be readily viewed.

Elodea

1. Using fingers, gently remove a healthy-looking leaf, those near the tip of the plant are often best. Place it on a clean slide, add one drop of water and place a coverslip over the leaf.
2. Locate a typical cell and draw a picture of it on the next page.
3. Remove the water from under the coverslip by holding a piece of paper towel at the edge of the coverslip. When most of the water has been removed, introduce several drops of a 10% NaCl solution by holding a pipette at the edge of the coverslip. Let it sit for 30 seconds and then observe under the microscope.
4. Draw one or two cells in which a noticeable change occurs. (Put the drawing on page 26)

Part 3. Concentration and Osmolarity

Concentrations of substances in clinical physiology are often expressed in units of mass percent in a solution. Water, the solvent for physiological solutions, has a mass of one gram per milliliter, so percent concentration refers to grams of solute per 100ml (100 g) of water. For example, a 0.9% NaCl solution contains 0.9g NaCl per 100 ml or 9g per liter. This is considered an isotonic saline solution since it has approximately the same concentration of osmotically active solutes (Na^+ and Cl^- ions) as normal body fluids.

Osmosis, the passive movement of water across a semi–permeable membrane, depends on differences in the total molar concentration of non–permeable (osmotically active) solutes on the two sides of the membrane. The total molar concentration of all solute particles in a solution is referred to as the **osmolarity** (osmoles per liter). The osmolarity of the intracellular and extracellular fluids is 0.3 osm/L (0.3 osmoles per liter) or 300 mosm/L (300 milliosmoles per liter).

To determine the osmolarity of a solution, you need to know the molar concentration of all the solutes in that solution. Note that a salt such as NaCl dissociates into two ions, so the osmolarity of a NaCl solution is two times the molarity of the undissociated salt. Concentrations are often expressed in grams, not moles, per unit of the substance (= number of grams per mole). The molecular weight is the sum of the atomic weights of all the atoms in the molecule.

Example 1: Calculate the number of moles in 100g of water.

H_2O: atomic weight of H = 1, O = 16; molecular weight of H_2O = 1 + 1 + 16 = 18

$100g\ H_2O \times 1mol/18g = 5.6mol\ H_2O$

Example 2: Calculate the molarity of ethanol (C_2H_5OH) in a bottle of beer containing 3% ethanol by weight.

atomic weight of C = 12, H=1, O = 16; molecular weight of ethanol = $(2\times12)+(6\times1)+16=46$

$$3\%\ \text{ethanol} = \frac{3g}{100ml} = \frac{30g}{liter} \qquad 30g/L \times 1mol/46g = 0.65mol/L$$

Example 3: Calculate the osmolarity (Osm) of a 2 molar (M) solution of NaCl. Use this equation relating osmolarity and molarity.

Osmolarity = Molarity $\times$ # solute particles the molecule dissociates into in water

2M NaCl $\times$ 2 (NaCl dissociates into 2 solute particles in water) = 4 Osm NaCl

Elodea Drawing

Human Physiology Lab 5 Name ______________________

Osmosis

1. Record your data from the experiment with an artificial membrane in table below.

Tme, minutes	0	15	30	45	60
weight of bag, g					

What conclusion did you draw from your data?
Which solution (water, 15% sucrose, 30% sucrose) was in your bag? ______________
Which of these solutions was in your beaker? ______________

2. Red blood cells ______________ in a hypertonic solution.

A 0.10 M NaCl solution is ________ (iso/hypo/hypertonic) relative to a 0.10 M glucose solution.

3. Concentration and Osmolarity

Show that a 0.9% NaCl solution is isotonic with normal body fluids, 300 mosm.
(Atomic weight of Na = 23 & Cl = 35.4)

Calculate the osmolarity of a 5% glucose (or dextrose) solution.

4. Picture a beaker filled with a 300 mosm (isotonic to body fluids) solution. Name two ways you could make the solution in that beaker hypertonic to body fluids.

5. Before the invention of refrigerators, pioneers preserved meat by salting it. How might this work?

Laboratory Exercise 6

Nerve Stimulation

Introduction

The classic experiment that demonstrates key aspects of neuronal physiology involves pithing frogs, isolating the sciatic nerve, and watching action potentials on an oscilloscope screen. The good news is that since many people have attained a higher level of ecological awareness, it is no longer acceptable to kill frogs in order to educate undergraduate physiology students. The bad news is that it is still acceptable to use electrodes on physiology students in order to educate them. Therefore, in this lab, you will be stimulating the nerves of your lab partner! (This is not the same as getting on their nerves.)

We cannot record action potentials from human nerves without performing surgery, you will therefore not be looking at action potentials. However, nerves (and muscles) may respond to *external* electrical stimuli because their cells conduct action potentials, which are electrical phenomena. You will stimulate various regions of your lab partner's body with mild electrical stimulation, and note when the stimulation is felt. This is a very crude approximation of the infamous frog sciatic nerve experiment. In the latter, the nerve is touched with electrodes to initiate action potentials, and the potentials are recorded with an oscilloscope. In our experiment, voltage is applied to the skin, and your brain, noting that a tingling sensation is felt, substitutes for the oscilloscope.

You will also examine the concept of threshold, which is an integral part of understanding how action potentials are generated in the body. **Threshold** is a general term meaning the minimum stimulus that just produces a response. In the context of a single neuron, there is a specific threshold (~ –50mV) at which an action potential is initiated. Graded potentials must usually be summed to reach this value. In our experiment, the subject also requires a particular level of stimulus to be able to feel something. This level (or threshold) will be measured with a single stimulus and then multiple stimuli. We expect that the threshold will be reached at a lower voltage if the frequency of stimulation is increased.

Anatomy note: you will record threshold voltages in three different parts of your body. The three parts have relevant antomical differences. Fingers are covered with thick skin, the epidermis is made of keratinized stratified squamous epithelium. Wrists are covered with the same epidermis, but thin skin. The tongue is covered with non–keratinized stratified squamous epithelium, and is coated with saliva, a watery electrolyte solution. This information is relevant for forming **hypotheses**, which you might be asked to generate after you have made **observations** about your data.

It is strongly recommended that you use the Interactive Physiology CD ROMs (in the lab and HLRC Center) to gain a more active insight into action potentials than can be provided in lecture or the text.

Objectives

- Become familiar with use of the electronic stimulator.
- Measure threshold for detecting stimulation with electrodes.
- Observe variations in sensitivity between body regions and individuals.
- Be able to graph and interpret class results.

Materials

Electronic stimulator
Electrodes and electrode gel
Alcohol swabs

Procedures

Become familiar with all controls on the **stimulator box** (Figure 6.1) and these terms: frequency, duration, volts. Facing the control panel, the MODE switch is on the left–hand side. The frequency is controlled using the knob and multiplier switch labeled EVENTS/SEC. The next knob to the right (DURATION-MS) controls the duration of the stimulus in milliseconds (ms). This knob should be set at 20 ms and the multiplier switch should be set at 1x. DO NOT change the duration settings at any time during the experiment. The next knob to the right is labeled AMPLITUDE–VOLTS (X.1) and controls the voltage. Note that the values shown are x 0.1, so that when the knob shows a value of 500 it is delivering 50 volts. The switch farthest on the right is the POWER switch.

Figure 6.1–Simulator Box–Image courtesy of Phipps & Bird

Everyone does this lab; each person tests the tip of their left little finger, their left wrist and the tip of their tongue. We are measuring **threshold** (not pain!). Record the value when the subject **first** detects the stimulus or a change in sensation. Most people will feel a slight tingling, some people feel a pulse or heat. Threshold values for finger and wrist usually start in the 30 volt range; for the tongue around 0.5 volts. Some people never feel anything on finger or wrist.

Attach electrodes to the STIMULUS OUTPUT (red with red and black with black). Clean the skin and wrist with alcohol pads and apply a thin layer of electrode gel before stimulation. For the tongue do not use alcohol pads or the electrode gel. Moisten a kimwipe with Quat 64 and use this to disinfect the electrodes before placing them on the tongue.

Single Stimulus

Experimenter will deliver a single stimulus. Set the MODE switch to OFF. In this setting the frequency controls are inoperative. Depress and release the SINGLE switch. Increase the voltage and depress the SINGLE switch again. Repeat this process until the subject first feels the stimulus. For finger and wrist, start at 10 volts and increase the stimulus in 5–10 volt increments until a definite current is felt. For the tongue, start at 0.5 volts and increase stimulus in 0.5 volt increments until a definite current is felt. Record this as the THRESHOLD voltage. Record your own data in your lab report and in the data table for the entire class; values greater than 100 volts (i.e. nothing was felt) should be recorded as > 100 volts – do not leave the table blank.

Multiple Stimuli

Repeat experiment starting with the threshold voltage determined above for each skin region; but this time giving multiple stimuli. Set the MODE switch to CONTINUOUS. The frequency knob (EVENTS/SEC) is now operative and the stimulator will deliver stimuli at the prescribed frequency. You no longer need to use the SINGLE switch.

1 stimulus/second: adjust the EVENTS/SEC knob to 1 and the MULTIPLIER switch to x 1. Turn the AMPLITUDE knob until the subject first feels the stimulus.

5 stimuli/second: adjust the EVENTS/SEC knob to 5 and the MULTIPLIER switch to x 1. Turn the AMPLITUDE knob until the subject first feels the stimulus.

25 stimuli/second: adjust the EVENTS/SEC knob to 2.5 and the MULTIPLIER switch to x 10. Turn the AMPLITUDE knob until the subject first feels the stimulus.

Record your own data in your lab report and in the data table for the entire class; values greater than 100 volts (i.e. nothing was felt) should be recorded as >100 volts – do not leave table blank!

Note: The student guinea pig has control at all times. If a stimulus becomes uncomfortable, they can just let go of the electrodes!

Human Physiology Lab 6 Name ______________________

Nerve Stimulation

1. Personal Data

Threshold, volts

	Single	Multiple Stimuli		
	Stimulus	1/sec	5/sec	25/sec
Finger				
Wrist				
Tongue				

2. Define the term threshold.

   In this experiment what do you *expect* to happen to threshold as frequency of stimulation increases?

3. Name two variables in this experiment that could be controlled?

   Name two variables that were recognized but could not be controlled?

4. You will be provided a graph of class data for this experiment. What **observations** can you make based on the data collected in this experiment? (Note – answer this after receiving class graph.)

5. After receiving class data, propose a **hypothesis** that might explain each of your observations. Do this as you prepare for the next lab exam.

Laboratory Exercise 7

CNS Anatomy

Introduction

This is a very simple, straightforward lab. It is actually an anatomy lab. In this section of the course we are discussing the CNS, and it is difficult to follow the discussion without having some idea of the location of the various key regions of the brain and their three dimensional relationships. The complexity of the nervous system resides primarily in its "wiring diagram". The CNS wiring is organized in "pathways", which consist of a chain of two or three neurons. The cell bodies of these neurons reside in specific gray regions (cortex, nuclei, spinal cord horns) and the axons contribute to specific tracts in the white matter of the CNS. Knowing these pathways is not only important for understanding how the brain works, it is critical for neurological diagnosis. Very simple, non–invasive procedures can help clinicians diagnose where an injury, infection, or vascular problem may be located in the CNS, if they are familiar with these pathways.

In this lab human brains, models and a neuroanatomy atlas will be reviewed so that specific parts of the brain, spinal cord, and neurons can be identified. Pictures of a variety of neural pathways can be viewed but you are not responsible for learning these.

Objectives

- Learn names & functions of major parts of the brain, spinal cord, neurons: list on pg. 35.
- Recognize these parts on models.
- Be able to distinguish the following:
  white & gray matter; neuron & nerve; nerve & tract; horn & column; cortex & nuclei

Materials

Brain, spinal cord, and neuron models

Procedures

Observation & reading text!

 Name ______________________________

CNS Anatomy

Be able to identify each of these structures on models, as well as answering the questions asked.

BRAIN: Give a function for each of these structures.

Cerebral cortex

1. Frontal Lobe: ______________________________

   Pre–central gyrus: ______________________________

2. Parietal Lobe: ______________________________

   Post–central gyrus; ______________________________

3. Temporal Lobe: ______________________________

4. Occipital Lobe: ______________________________

5. Insula: ______________________________

Cerebellum: ______________________________

Diencephalon:

1. Thalamus: ______________________________

2. Hypothalamus: ______________________________

Brain Stem:

1. Midbrain: ______________________________

2. Pons: ______________________________

3. Medulla oblongata: ______________________________

SPINAL CORD: Indicate the specific neuron structures found in each of these structures.

1. Gray horns: ______________________________

2. White columns: ______________________________

3. Dorsal root: ______________________________

4. dorsal root ganglia: ______________________________

5. Ventral root: ______________________________

6. Spinal nerve: ______________________________

NEURON: Give a function related to transmission of information for each of these structures:

1. Dendrite: ______________________________

2. Cell body: ______________________________

3. Axon: ______________________________

Laboratory Exercise 8

Reflex Arc

Introduction

The main function of the nervous system is communication, and this is carried out by electrical (and chemical) signals transmitted over very long cells (neurons) which function in part as do the wires in our home electrical system. Given this analogy, we can discuss the "wiring diagram" of the nervous system. In two other laboratory exercises (6 and 7) on the nervous system we have focused directly or indirectly on the concept that the nervous system is electrical in nature and has a specific wiring diagram. In this laboratory exercise we will focus on the most fundamental circuit in the nervous system, the reflex arc. A **reflex** is a relatively simple motor response to a particular stimulus. A **reflex arc** is the circuit that supports or makes possible the reflex action. A reflex arc includes a sensor, sensory neuron, interneuron (not present in all reflex arcs), motor neuron, and effector (muscle or gland). Since a specific simple reflex arc is associated with specific spinal cord segments and involves specific nerves, tests for reflexes are very useful in diagnosing neurological disorders.

We are going to focus on one particular reflex, the knee–jerk reflex. There are several important things to note about this reflex: it is a **stretch reflex**, the sensory receptor involved is the muscle spindle, and it is a monosynaptic reflex arc. Stretch reflexes protect muscles from over stretch and play a role in subconscious postural adjustments. We will test for this reflex and note whether the expected response, muscle contraction, is observed. We will also compare this reflex with a more complex neural pathway in terms of **latent period**. The latent period of a reflex is the time between the stimulus and the onset of the effector muscle contraction. The latent period reflects the time it takes for the neural message to go from sensor to effector. * Because the speed of action potential conduction is quite fast, the *latent period primarily reflects the number of synapses in a particular pathway.* Synaptic transmission is a much slower event than action potential propagation. You will elicit an involuntary knee–jerk reflex in your subject. Measure the latent period, and then compare that value with the latent period obtained with a voluntary response involving cognition.

You will also learn about the **plantar reflex** because it is a very important neurological test. The plantar reflex is elicited by cutaneous receptors on the sole of the foot. In normal individuals stimulation of these receptors results in adduction and plantar flexion of the toes. Sensory information ascends from the lumbar region of the spinal cord to the sensory cortex of the brain. The reflex action then requires uninterrupted conduction of nerve impulses along pyramidal motor tracts, which descend directly from the cerebral cortex to lower motor neurons in the spinal cord. Damage anywhere along the pyramidal motor tract produces an abnormal response, the Babinski reflex. The toes will abduct and dorsiflex. Infants exhibit the Babinski's sign or reflex normally because neural control is not yet fully developed.

Note that we are studying the stretch reflex to learn about neural pathways in the nervous system, but this reflex is also very important in the muscle system. The sensor is the muscle spindle receptor; it detects the length of a muscle, or how much a muscle has been stretched. Review information about this reflex in your text in the chapter on muscles.

Objectives

- Describe the neurological pathway involved in a simple reflex arc.
- Compare this pathway with a more complex neurological pathway.
- Explain the meaning of a latent period.
- Explain the difference in length of latent periods observed in involuntary reflex vs. voluntary action.
- Describe the structure and function of muscle spindles.
- Explain the clinical significance of tests for simple reflexes.
- Be able to elicit the plantar reflex and understand its clinical significance.

Materials

Rubber mallet
Goniometer
Wired hammer
Computer with Flexicomp program
Blunt probe

Procedures

1. Have the subject sit at the edge of a table with their legs dangling freely.
2. Attach the transducer (goniometer) to their right thigh and leg, with the transducer box facing outward and the hinge aligned with the knee joint. (see Fig 8.1)
3. The computer must be set up to receive data, directions for this will be given by the lab instructor.
4. Strike the subject's patellar ligament with the wired mallet. Practice eliciting this reflex a few times before collecting data. Repeat the experiment several times until you obtain a good trace.
5. Repeat the experiment to obtain the voluntary response data. This time the subject is to kick their leg when they hear the sound produced by the computer.
6. Measure and record the latent periods for these two experiments.
7. Review the anatomy and function of muscle spindles.

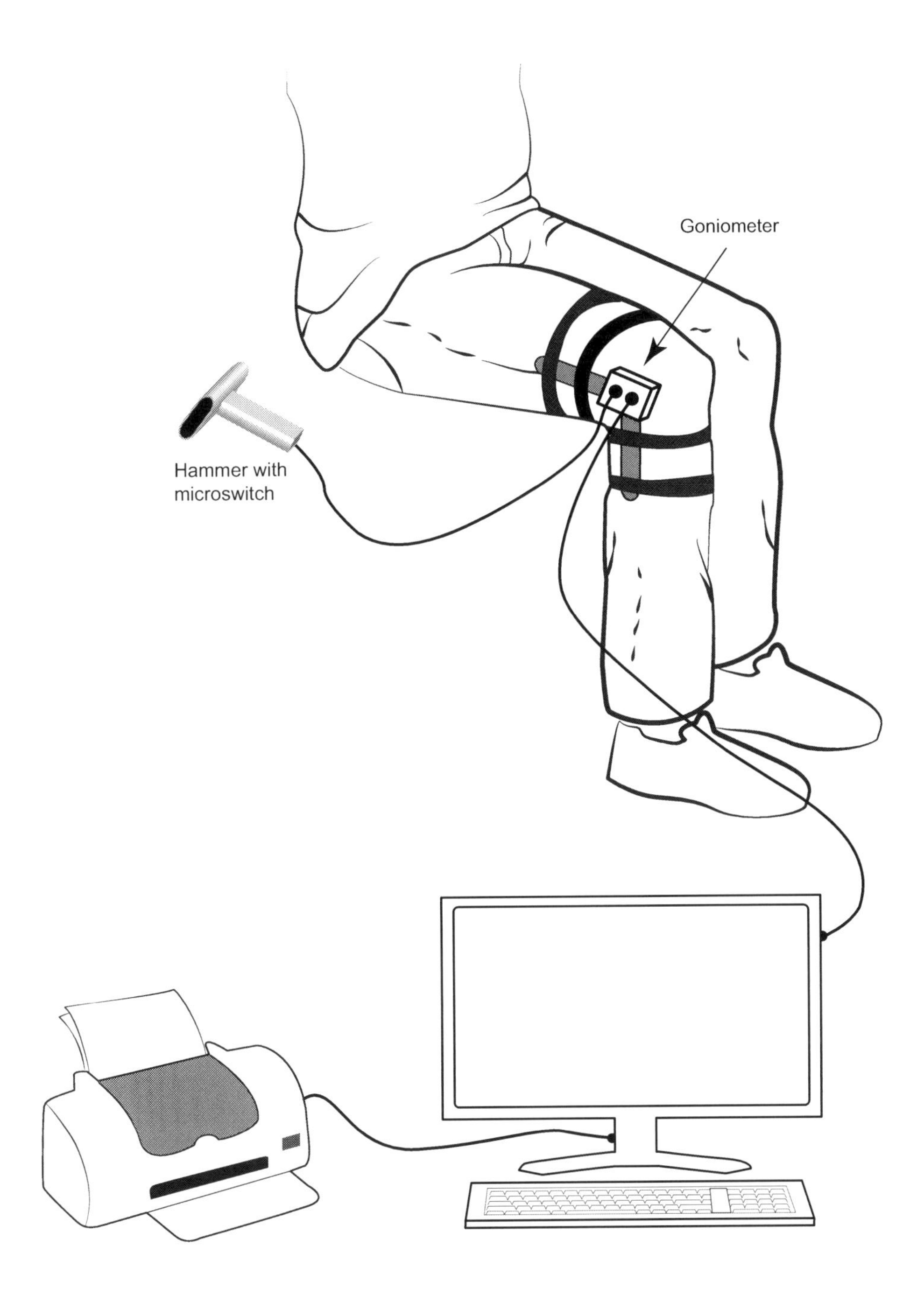

Figure 8.1 – The Flexicomp system arrangement of equipment

Human Physiology Lab 8 Name ______________________________

Reflex Arc

1. Record your data:

	Involuntary Reflex	Voluntary Response
Latent Period (sec)		

Compare the latent periods from several trials for both involuntary reflex and voluntary action. Which latent period was longer? ______________________________

2. Label the latent period in this data record and define the term latent period:

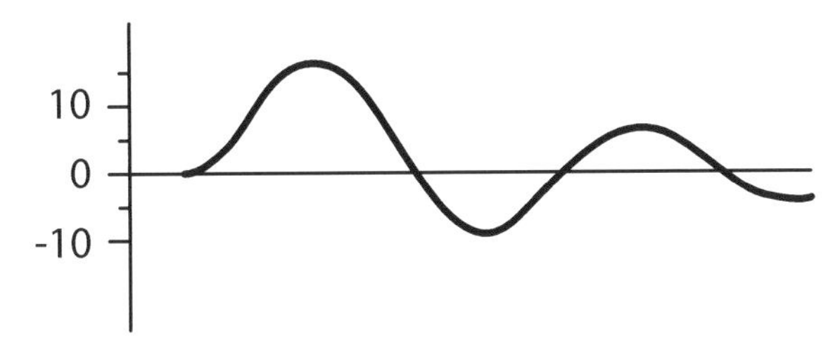

3. Describe the major reason for the difference noted between the latent periods of the involuntary reflex and the voluntary action. Answer should reflect information about the physiology of neural pathways.

4. List in order component parts of the knee jerk reflex arc, include specific name for sensory receptor:

_______________ _______________ _______________ _______________

Name the 5th component part found in most reflex arcs but absent above: _______________.

5. Think about the neural pathway for a stretch reflex (see #4 above) and the neural pathway for the plantar reflex (read introduction to this lab). Explain why damage to the spinal cord at the cervical level would affect the plantar reflex.

How would it affect the knee–jerk reflex?

Laboratory Exercise 9

Senses

Introduction

In order for the body to react in a purposeful manner to changes in the external and internal environment, the CNS needs information concerning the nature of the environmental change. Such information is generated by specialized structures at the beginning of sensory nerve fibers. These structures receive stimuli (changes in the environment) and are therefore called **sensory receptors**. They then change the incoming signal into a train of action potentials that are sent to specific parts of the brain for interpretation.

There are so many kinds of sensory receptors that they are usually organized into subcategories. The primary ones are **general senses**, which are found body wide, and **special senses**, localized in the head. General senses are further categorized as somesthetic – information about body feelings (touch pressure, temperature, pain) and **proprioceptive** – information about body position (muscle stretch and contraction, joint position).

In this lab you will learn about some of the *general properties* that most sensory receptors share, and perform experiments involving two major senses: *hearing* and *vision*. The general ideas about sensory receptors that we will study include: specificity, perception, acuity and adaptation. All sensory receptors detect some kind of stimulus, a form of energy, and transduce this signal into action potentials which can be transmitted to the brain for interpretation. Each sensory receptor is very **specific** and responds most readily to only one form of energy, the **adequate stimulus**. Thus, regardless of how a sensory receptor is stimulated, any given receptor gives rise to only one sensation. This observation is known as the Law of Specific Nerve Energies. The fact that sensory **perception** is a function of the brain suggests that perception may not always match objective reality. **Acuity** is discriminative ability, and refers to the ability of sensory receptors to discriminate between two different points of stimulation. **Adaptation** refers to the ability that some sensory receptors have to reduce the signal sent in response to a continued stimulus.

Hearing is normally produced by the vibration of the oval window in response to sound waves conducted through the movements of middle ear ossicles. However, the endolymph in the inner ear can also be made to vibrate in response to sound waves conducted through the skull bones, which bypass the middle ear. This makes it possible to differentiate between deafness resulting from middle ear damage (**conduction deafness**, e.g. from otosclerosis) and deafness resulting from inner ear damage to the cochlea or damage to the vestibulocochlear nerve (**sensory deafness**, e.g. from infections, prolonged exposure to loud sounds). The Rinne test is used to make this differentiation, the Weber test confirms the result.

In the sense of vision photoreceptors, rods and cones, transduce light energy into action potentials. The cornea and the lens of the eye together function to focus the light rays on the photoreceptors of the retina. The sensations of depth, movement, and color are creations of the brain, which receives, modifies, amplifies, and interprets the action potentials from the retina. The various tests that you perform in this lab will provide a better understanding of the physiological function of each of the key structures in the eye. Review these structures using eye models.

The primary function of the eye, exclusive of the retina, is refraction (bending) of the light rays so that they are focused on the retina. The cornea and lens perform this function. When light rays that diverge from two adjacent points in the visual field are each brought to a perfect focus on the retina, two points will clearly be perceived. This property is termed visual acuity, the power to discriminate details. If the light rays converge in front of or behind the retina, the two points will be perceived as one fuzzy or blurred point. This occurs if the shape of the eyeball is elongated or shortened, and is termed myopia (nearsightedness) and hyperopia (farsightedness) respectively. If there is an uneven curvature of the cornea and/or lens, there is also a loss of visual acuity. This is termed astigmatism.

The role of the *lens* in refracting light is special in that the shape of the lens can change so that objects at different distances can be brought into focus. A camera can be adjusted to focus on objects at varying distances by moving the lens, in the eye the shape of the lens is changed. This is accomplished by contraction of the ciliary muscle. This ability of the eye to focus the images of objects that are at different distances is called **accommodation.** As we age the lens becomes less elastic and we suffer a loss of accommodation. We hold the newspaper further and further away or get bifocals!

Light is admitted into the eye through an adjustable aperture, the *pupil*, surrounded by the *iris*. The iris consists of two groups of muscles with opposing actions, one constricts the pupil in bright light, the other dilates it in dim light. These muscles are reflexively controlled by the autonomic nervous system.

The *retina* is a thin layer of neural tissue that lines the back of the eyeball and contains sensory receptor cells, the rods and cones, as well as several types of neurons. Some visual processing is done in the retina, even before nerve impulses are sent to the thalamus and occipital cortex. A test that reveals the "**blind spot**" demonstrates that you have a small region of the retina where there are no receptors, and therefore no image can be formed when light is focused on this area. The retina contains a pigment which dissociates when it absorbs and transduces light energy. After this pigment has been bleached or dissociated by light, it has to be regenerated before the reaction can occur again. This aspect of vision physiology can be seen by testing for the **afterimage**. Cones are specialized for **color vision**. There are three kinds of cones, each responding maximally to wavelengths of light in the red, green, and blue regions of the visible spectrum. When one or more of these cones is defective, color blindness results. Looking at specially designed color plates can reveal if one has inherited a genetic defect that produces color blindness.

Objectives

- Demonstrate the Law of Specific Nerve Energies.
- Demonstrate a difference between sensation and perception.
- Demonstrate and explain sensory adaptation.
- Determine two point touch threshold in various areas of skin, explain differences noted.
- Perform two standard hearing tests, discriminate between sensory and conduction deafness.
- Learn anatomy of eye: cornea, iris, pupil, lens, retina, fovea centralis, optic disc
- Learn anatomy of ear: tympanic membrane, ossicles, cochlea, semicircular ducts, ampulla, vestibule, utricle, saccule, round window, oval window
- Perform tests for visual acuity, astigmatism, accommodation, color blindness, after image.
- Demonstrate presence of blind spot, explain why light focused here can't be seen.

Materials

Gloves, beaker of water, calipers, coins, rubber bands
Tuning forks
Ear & eye models
Eye charts: Snellen, astigmatism, color vision
Colored squares on black paper; cards with + and dot; meter sticks; pins
Penlights

Procedures

Specificity: Law of Specific Nerve Energies

Specific types of receptors are more sensitive to certain types of stimuli (e.g. the eye to light, taste buds to chemical stimuli). The type of stimulus to which the receptor is most sensitive is called the **adequate stimulus**. Nearly all receptors will respond to a stimulus other than the adequate stimulus if the stimulus strength is great enough. Cones of the retina may respond to mechanical stimulus even though their adequate stimulus is light: we "see stars" when hit in the head! Regardless of the type and strength of a stimulus applied to a given receptor, the sensation perceived when the receptor responds is always that sensation for which the receptor is designed. Regardless of the type of stimulus applied to the eye, the stimulus will always be perceived as light. This phenomenon is known as the law of specific nerve energies. Go to the computer room to conduct this experiment. In the darkened room, with your eyes closed, turn your eyes as far as possible to the left. Using your right index finger, gently press the outer part of the right eyeball. A dark circle surrounded by a bright white ring near the ridge of the nose will soon be "seen." This visual sensation of light is caused by stimulation of the rods and cones of the retina, receptors for vision, by mechanical pressure against the eyeball.

Name one type of stimuli in the environment for which you do not have sensory receptors, and therefor cannot detect? ____________________________

Perception

Cutaneous receptors are rarely stimulated singly. Usually the nature of stimuli is such that several different types of receptors are stimulated simultaneously. Cerebral association and interpretation of stimuli produce a broad spectrum of complex sensations from combinations of simple basic sensations. Place your hand in a surgical glove. Dip the hand into a beaker of 20 C water. Although the hand remains dry, the sensation of wetness is felt because its two components (temperature and pressure) are sensed.

What was the sensation with the gloved hand? ____________________________

Acuity: two point discrimination

The density of touch receptors is measured by the two point threshold test. Use calipers for this determination. Start with the calipers wide apart and the subject's eyes closed. The two points of adjustable calipers are simultaneously placed on the subject's skin with equal pressure, and the subject is asked whether two separate contacts are felt. If the answer is yes, the points are brought closer together, and the test is repeated until only one point of contact is felt. But note that it is important to randomly alternate two point contacts with one point contact so that the subject cannot guess the "correct" answer. The minimum distance at which two points can be felt is the two–point threshold. Determine the two point threshold on the palm, back of the hand, fingertip and back of the neck. The density of receptors or the **receptor field size** of any area of skin will be inversely related to the size of the somatosensory cortex devoted to that part of the body.

Adaptation: touch

When a stimulus of threshold strength or greater is applied to a receptor, the receptor discharges, resulting in a train of impulses conducted along the afferent neuron to the CNS. If the stimulus strength remains constant and the stimulus is not removed from the receptor, the receptor may become increasingly less sensitive to the presence of the stimulus. This phenomenon is known as adaptation. Some receptor types adapt rapidly to the presence of stimuli (e.g. Meissner's corpuscles to light touch), and others adapt very slowly (e.g. taste receptors) or not at all (e.g. cutaneous pain receptors.) Place a coin on your forearm, and let it remain for a few moments. The temperature of the coin must be near skin temperature (about 35½C), and the forearm must be kept motionless. How long does the sensation persist? Most cutaneous receptors (excepting pain receptors) adapt rapidly. If a stimulus to which the receptor has adapted is removed from the receptor, the receptor will again discharge, giving rise to the same sensation perceived when the stimulus was first applied, even though the original stimulus is no longer present. The sensation perceived after removal of the stimulus is known as the afterimage. Place a rubber band around your head, allow it to remain a few minutes, and then remove it. Compare the sensation perceived after its removal with that perceived on application of the rubber band, and note the duration of the "after– image."

What was the time to adaptation after placing coin on forearm? __________

What was the duration of the after image after removing rubberband? __________

Which sense adapt quickly? ______________________

Which senses adapt slowly, if at all? ______________________

Types of hearing loss:

Hearing loss can be characterized as either conduction deafness and/or sensorineural deafness.

Conduction deafness is caused by any impairment in either of the following:

1. **External auditory canal** – including impaction of ear wax in the auditory canal, damage to the tympanic membrane (*ruptured ear drum*), or narrowing of the external auditory canal (*stenosis*)

2. **Middle ear** – including otosclerosis (bone spurs on the ear ossicles) or otitis media (*inflamed and fluid-filled middle ear cavity*)

Sensorineural Deafness is caused by damage to either of the following:

1. **Inner ear** – damage to the hair cells resulting from
   - Chronic exposure to loud noise (live concerts, equipment)
   - Inflammation of the inner ear, tumor, trauma to the head, autoimmune disease, aging, and circulation problems.
   - Use of drugs that damage the inner ear including 1) Aspirin (when in large doses; 8 to 12 pills/day), 2) non-steroidal anti-inflammatory drugs (NSAIDs) such as ibuprofen, and naproxen, 3) some antibiotics such as streptomycin and neomycin, 4) loop diuretics used to treat high blood pressure and heart failure such as furosemide (Lasix) or bumetanide, and 5) medicines used to treat cancer.

2. **Neural pathways** – damage anywhere along the neural pathway from any cause.

Tests for hearing loss:

Hearing loss can be tested as described below using the **Rinne test** and the **Weber test**.

Rinne Test

Purpose: To detect conductive hearing loss

Instructions:

1. In a quiet room, strike a tuning fork and place the handle on the mastoid process behind the ear, with the tuning fork pointed down and back. Sound will be heard as it is transmitted to the ears through vibrations in the skull bones.

   Explanation: the vibrating skull bones make the fluids in the cochlea move. This causes the basilar membrane to vibrate which then bends the stereocilia into the tectorial membrane, creating action potentials that travel to the auditory cortex and sound is heard. Bypassing the inner ear and transmitting sound through the skull is called **bone conduction**.

2. When sound is no longer heard, place the tuning fork next to the external auditory meatus. The tuning fork will continue to vibrate and create sound waves.

   Explanation: If there is no damage to the middle or inner ear, these sound waves will follow along the normal pathway of sound as they pass through the auditory canal, vibrate the tympanic membrane, etc. Sound is heard through the normal pathway of **air conduction**.

   a. Usually **air conduction** is better than **bone conduction** so that a person with normal hearing will continue to hear sound when the tuning fork is moved from the mastoid process to the external auditory meatus.

   b. If the person does NOT hear the tuning fork when placed next to their external auditory meatus, this indicates **conductive deafness**. Causes are described above.

Weber Test

Purpose: To confirm whether hearing loss is conductive or sensorineural.

Instructions:

1. In a room with normal ambient noise, initiate **bone conduction** by striking a tuning fork and place its base at the center of the forehead. Determine if the sound is louder in the left ear, right ear, or equally loud in both ears.

2. **Explanation:**

   a. **Sound is equally loud in both ears** indicates the individual has **normal hearing**.

   b. **Sound is louder in the damaged ear** indicates **conductive hearing loss** as hearing in the damaged ear is "upregulated" (i.e. more sensitive to sound) relative to the normal ear.

   c. **Sound is louder in the normal ear** indicates **sensorineural hearing loss**. This is because even though the sound wave is conducted by bone to both ears, in the ear with sensorineural damage the cochlea has a diminished ability to generate action potentials.

Guide for Interpreting Weber and Rinne Tests

Condition	Weber Test	Rinne Test
no hearing loss	no lateralization (equally loud in both ears)	sound perceived longer by air conduction
conduction deafness	lateralization to deaf ear (louder in deaf ear)	sound perceived longer by bone conduction
nerve deafness	lateralization to normal ear (louder in normal ear)	sound perceived longer by air conduction in normal ear

Vision Tests

Visual Acuity & Astigmatism

Stand 20 feet from the Snellen eye chart. Remove glasses if you wear them, cover one eye, and attempt to read the line with the smallest letters you can see. Have your lab partner note which line you could see and read the chart to determine the visual acuity of that eye. Repeat for the other eye. The basis for the Snellen test is that letters of a certain size should be seen clearly at a specific distance by eyes that have normal acuity. For example, line 1 should be read easily at 200 feet, line 8 at 20 feet. A person's visual acuity is stated as V = d/D, in which d = the distance at which the person can read the letters and D is the distance at which a normal eye can read the letters. Stand 20 feet away from an astigmatism chart and cover one eye (glasses off). If an astigmatism is present, some of the spokes will appear sharp and dark, whereas others will appear blurred and lighter. Still covering the same eye, slowly walk up to the chart while observing the spokes. If you wear corrective lenses, repeat the tests with them on.

Visual acuity				Astigmatism (+ or –)			
right eye		left eye		right eye		left eye	

Accommodation

The near point is the closest distance at which one can see an object in sharp focus. You will need a meter stick and a pin. Take glasses off if you wear them. Hold the meter stick under one eye and extend it outward. Close the other eye. Hold the pin at arm's length away resting on the stick. Gradually move the pin forward. Record the distance at which the pin first appears blurred. This is the **near point of vision**. Record your data here and compare your data with normal values. We expect the numbers to increase with age.

age (years)	10	20	30	40	50	60
near point (cms)	7	10	14	22	40	100

Blind Spot

The axons of all ganglion cells (one cell type in the retina) gather together to become the optic nerve, which exits the eye at the optic disc. This is also called the blind spot because there are no rods or cones here, so an object whose image is focused here will not be seen. Test this by holding the card with the + and o symbol about 20 inches away from your face, with the left eye covered or closed. Keep the right eye focused on the circle, slowly bring the card closer to your face until the cross disappears. Continue to move the card towards your face until the cross reappears.

Afterimage

The light that strikes receptors of the eye stimulates a photochemical reaction: a photopigment dissociates. This chemical reaction produces electrical changes in the photoreceptors which trigger action potentials. These events can't be repeated in a given rod or cone until the pigment is regenerated. This involves another chemical reaction, involving vitamin A. Therefore a certain period of time is required before a receptor can again be stimulated. Test this by staring at a light bulb for a period of time and then shifting your gaze to a blank wall. The bright image of the bulb will still be seen. This is called the positive afterimage and is caused by the continued firing of the photoreceptors. After a short period, the dark image of the bulb will appear. This is the negative afterimage and is due to the bleaching of the visual pigment of the affected receptors.

A more interesting version of this test involves bleaching the pigment of the cones, the color receptors. Stare at one of the colored squares on black cardboard for 3 minutes with one eye only. Quickly shift your gaze to a sheet of white paper, and note the color of the positive afterimage. The cones that have been responding will be bleached. When you shift your gaze to white paper, all cones will be stimulated by the white light, but only the cones that are not bleached can respond, and you will see the complementary color. Perform this test with red, blue and yellow squares and record your results on the next page, number 5.

Color Vision

A series of color plates are observed which can reveal the different kinds of colorblindness. A sheet of paper with details on how to use and interpret this color vision test will be available in the lab.

Human Physiology Lab 9 Name ______________________________

Senses

1. Define "adequate stimulus": ____
   Indicate the adequate stimulus for each of the following sensory receptors:

   cones of retina ________________ Pacinian corpuscle ________________

   hair cells of cochlea ________________ Meissner's corpuscle ________________

2. The Rinne test can differentiate conduction & nerve deafness. Why do patients with otosclerosis hear better when the tuning fork is on bone rather than near the affected ear?

   In sensorineural impairments of hearing, the sound in Weber's test is heard better in the unaffected ear. Why?

3. Acuity – record your data for two point threshold in touch perception (in mm):

back of hand	palm of hand	fingertip	back of neck

   What do these data suggest about receptor field size and a map of the somatosensory cortex?

4. When testing for the blind spot, can it be sensed when both eyes are open during the test? Why or why not?

5. Identify the color of the positive afterimage seen with these colored squares & explain these findings:

   red ____________ blue ____________ yellow ____________

Laboratory Exercise 10

Muscle Contraction

Introduction

The main function and most unique characteristic of muscle is contraction. There are various ways to observe skeletal muscle contraction. Ask a classmate to rest a bare forearm on the lab bench and write a few words on some paper while you watch their arm. You will appreciate how muscles were named. Mus is part of the Latin word for mouse and people thought that moving muscles looked like mice scurrying under the skin. In our muscle lab we will study three aspects of contraction:

- the anatomy which supports this function
- the contraction of individual muscle cells and the role of ATP in this action
- the contraction of whole muscle and the initiation of this action by electrical stimulation

The anatomy of muscle fibers will be learned by using models. Be sure you can identify and differentiate the following structures: muscle fiber, myofibril, myofilaments, sarcomere, Z disc, A, I & H bands, transverse tubules, sarcoplasmic reticulum, neuromuscular junction (both axon terminal & motor end plate).

Individual muscle cells, obtained from rabbit muscle, can be stimulated to contract with ATP, and observed during contraction using the light microscope. In this lab we will also examine the contraction of whole muscles. You will use classmates as guinea pigs, and take advantage of the fact that contraction of muscle, which is normally initiated by nerve impulses, can also be stimulated by external electrical stimulation.

You will observe three properties of muscle contraction: twitch, summation and tetanus. When a single electrical stimulus is applied to a muscle, it quickly contracts and relaxes. This response is called a **twitch**. Increasing the stimulus *voltage* increases the strength of the twitch, up to a maximum. Increasing the *frequency* of the stimulus produces a second twitch before the first is complete, a response called **summation**. With progressively increasing frequency of stimulation at a set voltage, the time between twitches gets shorter and shorter, and the contraction increases in amplitude. Eventually there is no time between twitches and they fuse to form a sustained maximal contraction. This sustained maximal contraction is called **tetanus**.

Note that while summation of muscle twitches is possible, summation of action potentials is not. Because of the refractory period, action potentials are "all–or–none" events. Muscle twitches take much longer than action potentials and have no refractory period, hence summation is possible. Also remember from lecture that increasing the frequency of action potentials, which allows summation of twitches, is one way that **graded contractions** of muscles is made possible in the body. Recruitment of motor units is another way. **Sustained contractions**, whether of maximal strength or not, are achieved by means of a sustained stimulus (action potentials) and asynchronous recruitment of motor units to prevent fatigue.

Objectives

- Identify these structures on muscle fiber & myofibril models: sarcolemma, NMJ, motor end plate, myofibril; thick & thin filaments, glycogen granules, sarcomere, I, A, H bands, Z & M lines (discs)
- Observe the contraction of isolated muscle fibers; describe the role of ATP in this process.
- Describe how muscle contraction can be stimulated by a direct electrical stimulus
- Define and demonstrate twitch, summation and tetanus
- Explain how graded muscle contractions and sustained muscle contractions are normally produced.

Materials

Glycerinated rabbit skeletal muscle
ATP solution
Needles for teasing apart muscle fibers
Light and dissecting microscopes, slides
Microscope slides and coverslips
Muscle fiber and myofibril models
Stimulating Electrodes & stimulator
Plastic strip for anchoring the grounding electrode

Procedures

Part 1. Muscle Fiber Contraction

The muscle is isolated from rabbit and shipped in 50% glycerol, which prevents deterioration and desiccation. It will be placed in the back of the lab in small pieces (about 2 cm), and still in glycerol. Place the glass container with the rabbit muscle under the **dissecting microscope**. While looking through the dissecting microscope, use 2 sharp probes to tease apart the muscle segment to isolate thin groups of myofibers (muscle cells). If possible obtain single fibers as they will demonstrate the greatest contraction. Now you will transfer these fibers to slides as described below.

1. Use a bent probe to transfer five or fewer individual fibers (with glycerol) to each of 2 microscope slides. Position the fibers so that they are as straight as possible.
2. If fibers are not completely covered by glycerol, use the bent probe to transfer glycerol, from container with the rabbit muscle, onto the muscle fibers. It is important that fibers do not dry out.
3. **Slide #1**: cover the fibers with a coverslip and exam with your **light microscope** under low and high power. Can you identify individual fibers? Note the appearance of the fibers. Can you see striations?
4. **Slide #2**: do not cover the fibers with a coverslip. Be sure the fibers are kept moist with glycerol. Place the slide under the **dissecting microscope** and measure the length of your fibers with a millimeter ruler placed beneath the slide. What is the original length of your muscle fibers in milliliters? ______________
5. Next, flood the fibers on **Slide #2** with several drops of room temperature solution containing ATP plus potassium and magnesium ions. As you do this, observe the reaction of the fibers! After 30 seconds or more remeasure the length of the fibers. After adding ATP, what is the length of your muscle fibers in millimeters? ______________
6. In the questions at the end of this lab, you are asked to use the lengths measured above to calculate the degree of muscle contraction after the addition of ATP. Note that under favorable conditions, a myofiber can contract up to 50% of its starting length.

Part 2. Whole Muscle Contraction

Students will work in pairs using stimulating electrodes and the stimulator box. The electrode will be placed on the subject's arm, with electrolyte gel under the electrode. The electrode will be moved until the placement is such that a small stimulus will elicit contraction of only one finger.

Once a muscle twitch has been observed, set the stimulator so that it delivers one pulse of stimulation per second. Keep the stimulus intensity constant and gradually increase the frequency. Continue until tetanus and summation have been observed. Notice **twitch, summation** and **tetanus** in Fig 10.1 below.

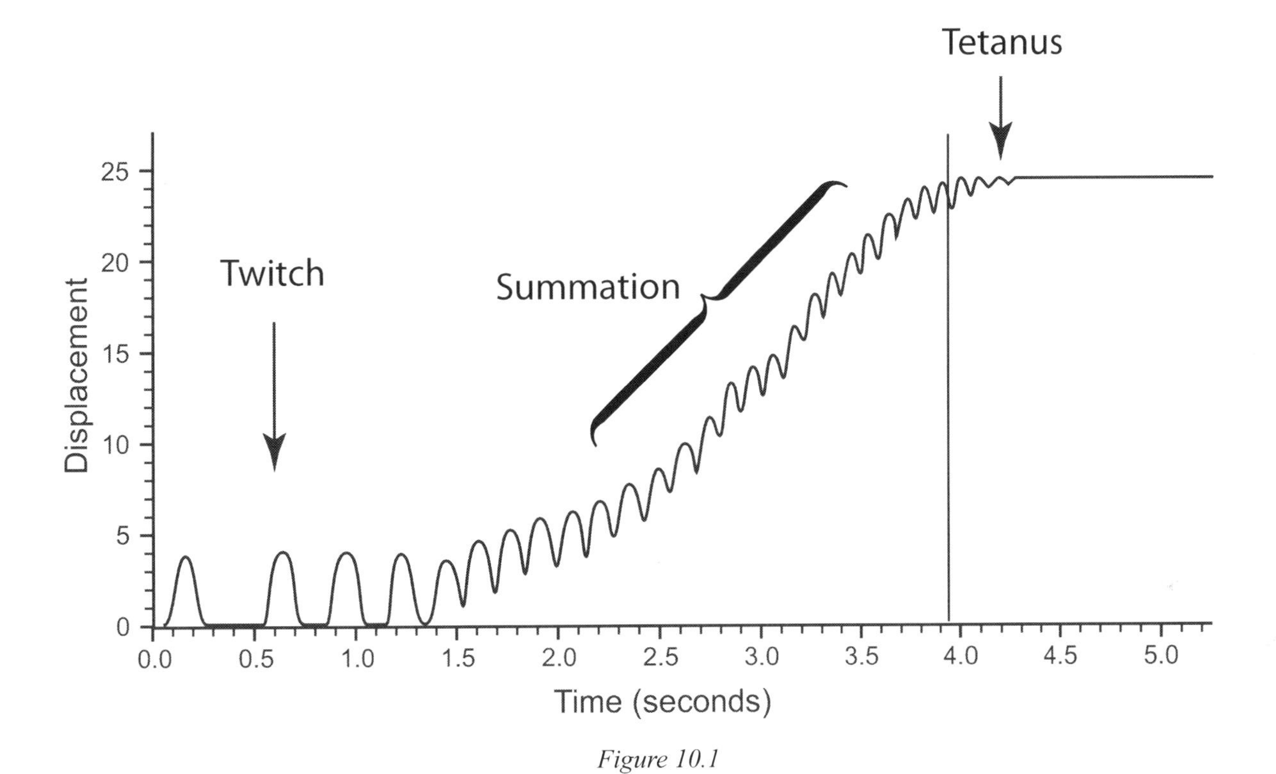

Figure 10.1

Human Physiology Lab 10 Name ___________________________

Muscle Contraction

1. Be sure you can identify the following parts of a myogram: twitch (contraction & relaxation period) summation, tetanus. Provide the following data for the whole muscle experiment:

   threshold stimulus: ______________ frequency which produced tetanus: ______________

   How was summation of muscle twitches produced in the lab? ___________________________

   How is summation produced normally in the body? ___________________________

2. Record the data you obtained with the rabbit muscle fibers:

	original muscle length	final muscle length	% contraction
ATP + salt			

3. Which band (A or I) can no longer be seen when skeletal muscle contracts? ______
   Why does this band "disappear"?

4. How are variations in the strength of muscle contraction produced (graded contractions)?

   How are sustained contractions produced in muscles?

5. Draw two adjacent relaxed sarcomeres and label all the parts.

Laboratory Exercise 11

Cardiac Function

Introduction

The heart is the pump for the cardiovascular system. The heart's ability to maintain adequate perfusion of the body's organs depends on proper electrical stimulation, muscle contraction, and functioning of the valves. We will assess two of these functions in this lab, by recording the electrical activity of the heart (an EKG) and listening to heart sounds, which are produced by the closing of the valves. This lab also provides an opportunity to review the anatomy of the heart, and the events of the cardiac cycle.

The opening and closing of the heart valves produce sounds that are audible over the ventral surface of the chest when correctly using a stethoscope. The first sound, usually verbalized as "lubb" is due to the closing of the atrioventricular valves. The second sound, verbalized as "dupp", is due to the closing of the semilunar valves. With careful and trained **auscultation** (listening) it is possible to detect various abnormalities of the heart. Valves that do not close properly produce gurgling sounds (**heart murmurs**) which are due to back flow of blood. Defects in the cardiac septum, the walls between the ventricles, or an open foramen ovale in the wall between the atria also produce detectable sounds as blood regurgitates between the chambers.

Unlike skeletal muscle, cardiac muscle can initiate its own contraction. All cardiac muscle cells can potentially initiate their own depolarization and contraction, but some cells have become specialized for this functionand depolarize at high rates. These rapidly depolarizing cells are called nodal cells; they initiate the waves of depolarization that pass through every cardiac cell, and initiate contractions of atria and ventricles. Since body fluids contain high concentrations of electrolytes, the electrical activity generated by the heart travels throughout the body and can easily be monitored by electrodes placed on different areas of the skin. A graphic representation of these electrical activities is called an electrocardiogram (**EKG** or **ECG**).

The sequence of electrical and mechanical events of the pumping of the heart is known as the **cardiac cycle**. With an experimental animal, an opened thoracic cavity, and more sophisticated equipment, one can observe and record all of the events associated with the cardiac cycle. Obviously, we are not going to perform this experiment. However, we can make a recording of an EKG on some human volunteers.

Objectives

- Identify the major anatomical structures of the heart, know their functions: atria, ventricles, AV valves, semilunar valves
- Listen to and describe the causes of heart sounds.
- Describe the normal EKG pattern and explain how it is produced.
- Obtain an EKG using the limb leads, identify the waves determine the duration of one cardiac cycle (i.e. one beat of the heart), and measure the cardiac rate.
- Become familiar with the concurrent events of the **cardiac cycle**: electrical events (EKG), pressure changes, valve openings and closings, and heart sounds, and be able to explain their relationships.

Materials

Heart models
Stethoscopes

Procedures

Anatomy

Using heart models be sure you can identify the following: atria, ventricles, right vs. left sides, atrioventricular valves (tricuspid and bicuspid), semilunar valves, interventricular septum. Be able to trace the path of a drop of blood through the heart, noting where there is oxygenated or deoxygenated blood. Notice the different design of the atrioventricular valves vs. the semilunar valves, and be able to describe the function of these two kinds of valves.

Heart Sounds

Listen for the heart sounds of your lab partner. Figure 11.1 shows a diagram of different places to listen to the sounds generated by the aortic semilunar valve, the pulmonary semilunar valve, the tricuspid valve (right atrioventricular valve), and the bicuspid valve (mitral or left atrioventricular valve). Each auscultatory area is good for detecting different heart valve defects. The bicuspid area is best for hearing the first heart sound, the aortic or pulmonic areas are best for the second heart sound.

EKG

The lab instructor will provide specific directions for obtaining the EKG recordings, using the Biopac computer equipment.

Label all the EKG waves on your own data recording and be able to relate the different EKG waves to the events of the cardiac cycle, as shown in Fig. 11.2.

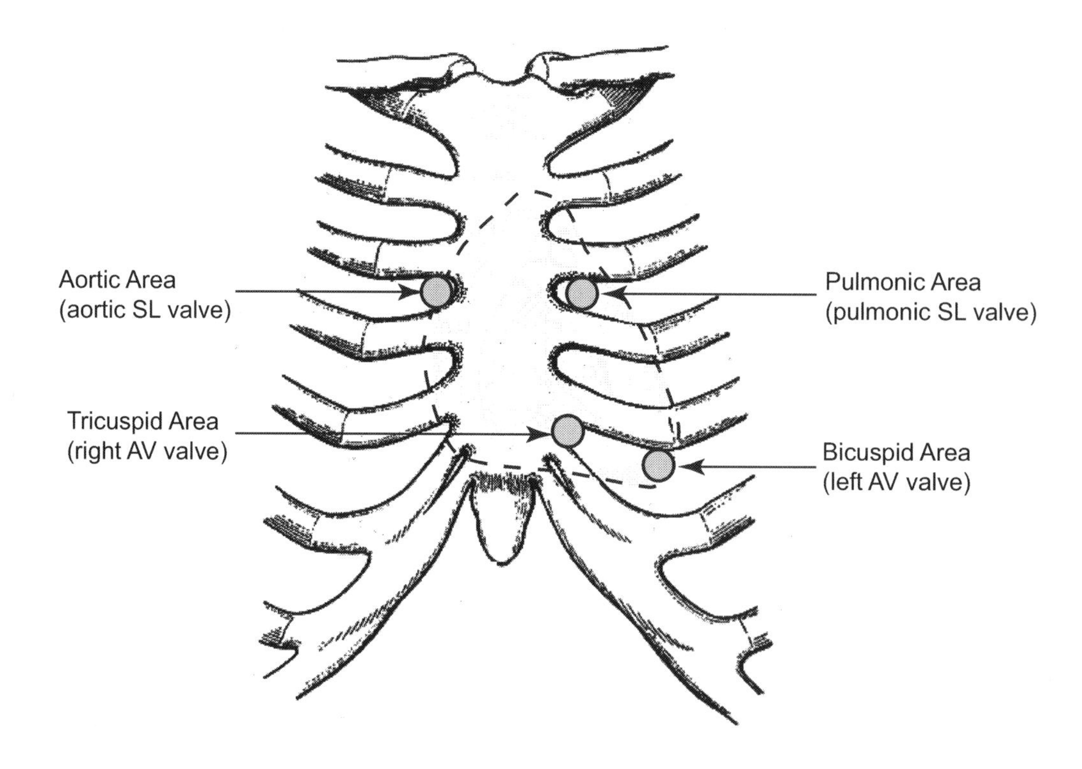

Figure 11.1– auscultatory areas for individual sounds

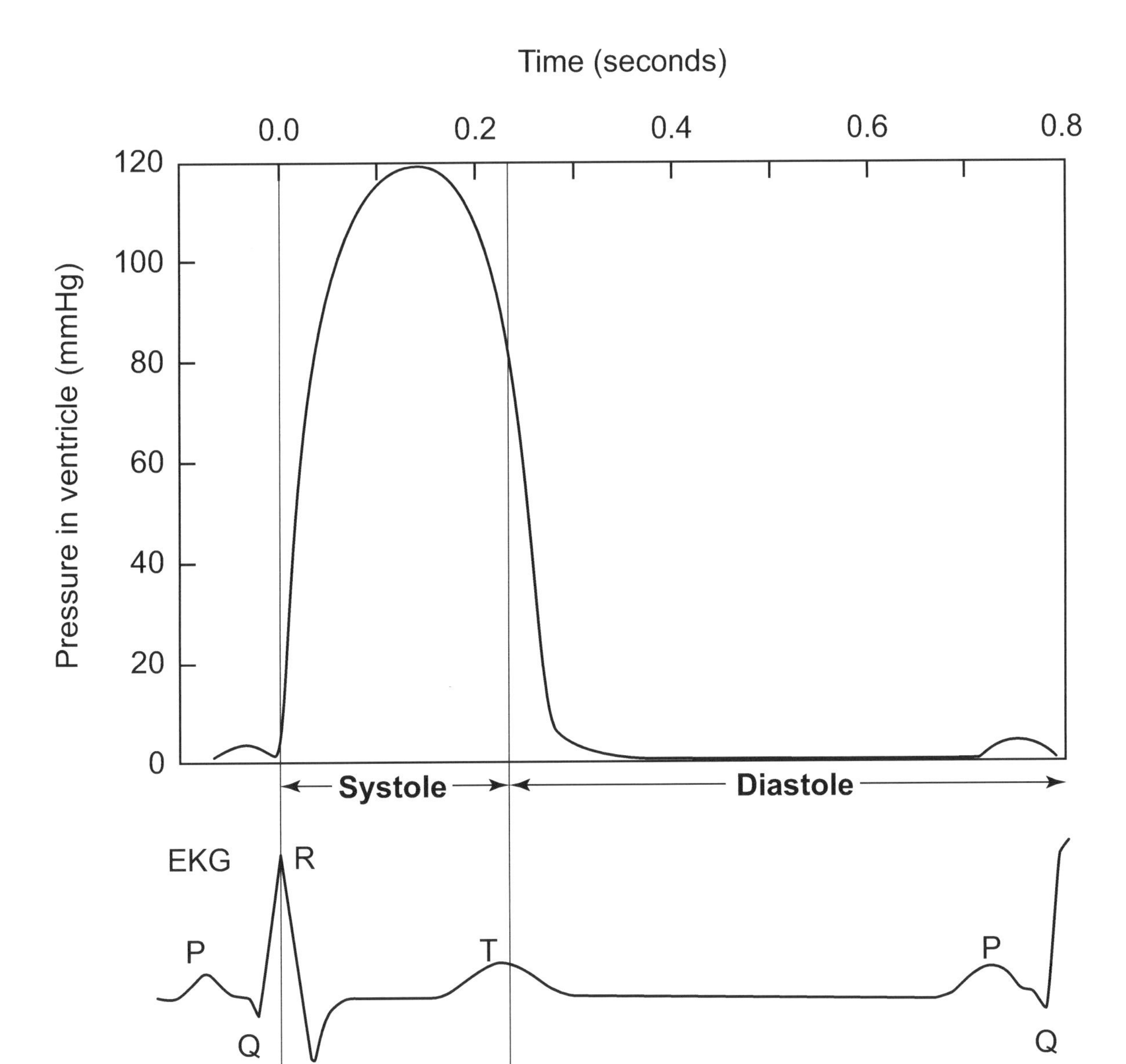

Figure 11.2 – Events of the Cardiac Cycle

Human Physiology Lab 11 Name ____________________

Cardiac Function

1. Be sure that you can label the events associated with an EKG recording. Name the electrical events that produce each of the following:

   P wave ____________________

   QRS wave ____________________

   T wave ____________________

2. What accounts for each sound heard when using the stethoscope to listen to heart?

   lubb ____________________ dupp ____________________

3. Explain why the SA node, rather than the AV node, functions as the normal pacemaker of the heart.

4. What is a heart murmur? Name two causes for this condition.

5. Explain why defects of heart valves can be detected by auscultation but not by electrocardiography.

6. Differentiate the following:

   Atherosclerosis ____________________

   Myocardial infarction ____________________

   Heart block ____________________

   Congestive heart failure ____________________

Laboratory Exercise 12

Blood Pressure

Introduction

Blood pressure is an essential parameter for the maintenance of homeostasis in the body, and is tightly regulated. Abnormally high or low blood pressures are a serious health risk, and its measurement is a routine part of health care. Blood pressure can be easily measured using a stethoscope and sphygmomanometer. In this lab you will practice measuring blood pressure. To observe how blood pressure and heart rate are affected by changes in posture, you will take blood pressure readings of a person in both reclining and standing positions.

Read in your text and reference lab manuals how the sphygmomanometer works. Basically it is a pressure cuff that occludes blood flow in an artery. You listen with a stethoscope for sounds of turbulent blood flow. Initially the artery is occluded and there is no flow (no sound). As you release the pressure cuff, blood flows during systole (heart contraction) but not diastole (heart relaxation). This produces turbulent flow and a sound, the first sound of Korotkoff. When the pressure in the cuff drops below diastolic pressure, smooth, laminar blood flow returns and the sounds stop. There are many more details about these sounds in reference books. The cuff pressure, read on the pressure gauge (manometer) in mm Hg, at which the first sound is heard is the **systolic pressure**. The pressure at which the sound disappears is taken as the measurement of **diastolic pressure**. There are two other measurements that you will be calculating. The **pulse pressure** is calculated as the difference in these two pressures, and when multiplied by 1.7 gives an estimate of stroke volume. The **mean arterial pressure** is equal to the diastolic pressure plus one–third of the pulse pressure. You will also measure **pulse rate** (heart rate).

You will measure blood pressure and pulse rate while reclining and immediately after standing. These measurements will provide insight into the **baroreceptor reflex**, which regulates blood pressure. You will also measure pulse rate before and after a brief period of exercise. This will allow students to evaluate their cardiovascular fitness level. A more complete evaluation of fitness would include tests of strength and flexibility.

Objectives

- Describe how sounds of Korotkoff are produced.
- Learn procedure to measure blood pressure.
- Correlate systolic and diastolic pressure with sounds of Korotkoff.
- Describe how pulse pressure and mean arterial pressure are calculated.
- Observe the effects of postural changes on blood pressure and pulse rate.
- Observe changes in cardiovascular function in response to exercise.

Materials

Sphygmomanometer
Stethoscope
Step bench

Procedures

1. Have the subject sit with his/her arm resting on a table at the level of the heart to take their blood pressure. The lab instructor and various experts among your classmates will make sure you learn how to make this measurement. Practice this procedure a few times before beginning the experiments.

   General instructions for taking blood pressure are as follows:
   Choose arm and move clothing, sleeves from upper arm.
   Put blood pressure cuff on subject's arm, about 1 inch above elbow crease.
   Diaphragm of stethoscope is placed over brachial artery, and not under pressure cuff.
   Ear pieces of stethoscope should be pointing forward.
   Inflate the pressure cuff until gauge reads about 180 mmHg, bulb valve should be closed.
   Release pressure by opening bulb valve slowly, pressure should fall smoothly.
   Listen for clicking or tapping sounds, note reading for first sound = systolic pressure.
   Listen for when sounds stop or are muffled = diastolic pressure.
   Once numbers are obtained, rapidly deflate the cuff.

2. The pulse rate is measured by placing your fingertips (not the thumb) on the radial artery in the ventrolateral region of the wrist. Count the number of pulses in 15 seconds and multiply by four.

3. The first set of measurements are made with the subject reclining, then immediately standing. Use a yoga mat on the floor for reclining. **Measure** the systolic and diastolic pressure, and the pulse rate first while reclining and record results in the data table on the last page of this lab. Then repeat these measurements immediately upon standing and record results. Calculate the pulse pressure and mean arterial blood pressure. Record these data in the data table on the last page of this lab and note which change of position affected the greatest change in each of these parameters. Next, use your data to calculate the parameters below and record your calculations in the same data table.

   pulse pressure = systolic pressure – diastolic pressure

   pulse pressure x 1.7 = stroke volume

   mean arterial blood pressure = diastolic pressure + 1/3 pulse pressure

   The baroreceptor reflex will be activated in a change of position from lying to standing. The parameter most directly affected by this reflex is the heart rate, measured as pulse rate. Usually the systolic pressure remains constant with changes in body position while the diastolic pressure may be affected by gravity.

4. Measure the pulse rate before and after a controlled period of exercise. The controlled exercise consists of placing your right foot on the lab exercise bench and stepping up and then down. Repeat this five times, allowing 3 seconds for each step up. Measure your pulse rate before, immediately after, and at 30, 60, 90, and 120 seconds after completing the exercise. Record these measurements on the last page of this lab.

5. Use data recorded on the last page of this lab (#1. & #2.) and Table 12.1 on the next page to calculate your fitness score. Individual tables in Table 12.1 are numbered; use the tables in this order. Use each table and your data to determine points earned per table. Totaling all points earned will give you a fitness score and a fitness rating that you can record on the last page of this lab (#3.)

1. Reclining Pulse Rate	
Rate	Points
50–60	3
61–70	3
71–80	2
81–90	1
91–100	0
101–110	–1

2. Standing Pulse Rate	
Rate	Points
60–70	3
71–80	3
81–90	2
91–100	1
101–110	1
111–120	0
121–130	0
131–140	–1

3. Pulse Rate Increse on Standing					
Reclining Pulse	0–10 Beats	11–18 Beats	19–26 Beats	27–34 Beats	35–43 Beats
50–60	3	3	2	1	0
61–70	3	2	1	0	–1
71–80	3	2	0	–1	–2
81–90	2	1	–1	–2	–3
91–100	1	0	–2	–3	–3
101–110	0	–1	–3	–3	–3

4. Change in Systolic Pressure from Reclining to standing	
Change (mmHg)	Points
Rise of 8 or more	3
Rise of 2–7	2
No rise	1
Fall of 2–5	0
Fall of 6 or more	–1

5. Pulse Rate Increase Immediately after Exercise					
Standing Pulse	0–10 Beats	11–18 Beats	19–26 Beats	27–34 Beats	35–43 Beats
60–70	3	3	2	1	0
71–80	3	2	1	0	–1
81–90	3	2	1	–1	–2
91–100	2	1	0	–2	–3
101–110	1	0	–1	–3	–3
111–120	1	–1	–2	–3	–3
121–130	0	–2	–3	–3	–3
131–140	0	–3	–3	–3	–3

6. Return of Pulse to Standing Normal after Exercise	
Seconds	Points
0–30	4
31–60	3
61–90	2
91–120	1
After 120	0

Your fitness score from all tables: ________________ total points
Evaluate your fitness score based on the Fitness Rating table below.
What is your Fitness Rating? ________________

Excellent	18–17
Good	16–14
Fair	13–8
Poor	7 or less

Table 12.1 – Tables for Calculating Fitness Score (Human Physiology Laboratory Guide, Fox)

Human Physiology Lab 12 Name ____________________

Blood Pressure

1. **Reclining and Standing Data** – below record your blood pressures, pulse rates, and calculated parameters:

	Reclining	Standing
systolic pressure, mmHg		
diastolic pressure, mmHg		
pulse pressure, mmHg		
mean arterial pressure, mmHg		
pulse rate, bpm		

2. **Exercise Data** – below record your *pulse rate* immediately before exercise, immediately after the specified exercise (time 0), and at 30 second intervals for 2 minutes.

		seconds post exercise				
	before	0	30	60	90	120
Pulse rate	______	______	______	______	______	______

Time for return of pulse rate to standing normal value: ______________

3. Record your fitness score and fitness rating below.

Fitness score: ______________ Fitness rating: ______________

4. There are changes in blood pressure when a person moves from a supine position to a standing position.

– What causes the initial change in blood pressure upon standing?

– What is the primary mechanism for the compensatory response to this initial change in blood pressure?

5. When blood pressure measurements are taken, the first sound of Korotkoff occurs when the cuff pressure equals the ____________________ and the cessation of sound occurs when the pressure equals the ____________________.

What causes the sounds of Korotkoff? ______________________________

6. Individual data should also be written in the class data sheet. Be prepared to make some conclusions about these data. Were there differences between athlete, non–athlete? (Note that for the purposes of this exercise an athlete is defined as someone who performs at least 30 continuous minutes of cardiovascular exercise three times a week).

Laboratory Exercise 13

Pulmonary Function

Introduction

The term **respiration** is applied to three separate but related functions: ventilation (breathing), gas exchange, which occurs between air and blood in the lungs and between blood and extracellular fluid (ECF), and oxygen utilization by cells in the energy liberating reactions of *cell respiration*. The volumes of air that move through the lungs during inspiration and expiration are termed respiratory volumes and capacities (see Table 13.1). These volumes can be measured or extrapolated using an instrument called a **spirometer,** and can be indicative of the pulmonary function of an individual. Many factors may contribute to one's pulmonary function, in both negative and positive ways. For example, exercise increases the **vital capacity** of a person and smoking will decrease vital capacity.

The ability to ventilate the lungs in a given amount of time is another important indicator of pulmonary function, and the **forced expiratory volume (FEV_1)** is such a measure. FEV_1 is defined as the volume of air that can be exhaled in the first second of a forced exhalation. When FEV_1 is expressed as a percentage of forced vital capacity, also known as percent FEV_1 (FEV_1/FVC x 100)), this percentage can be compared to normal values for healthy adults and is an indicator of airway obstruction. For example, a percent FEV_1 of 80% is considered normal for ages 18-29. (See table 13.2)

Chronic pulmonary dysfunctions can be divided into two categories: **obstructive** and **restrictive** disorders. These can be distinguished, in part, by the spirometry tests performed in this exercise. In obstructive disorders (emphysema, bronchitis, asthma) the airways are partially occluded by bronchiolar secretions, inflammation, bronchiolar smooth muscle contraction, and the FEV_1 will be low. In restrictive disorders (also emphysema, pulmonary fibrosis) the airways can be clear but there is lung tissue damage, and the vital capacity will be reduced.

Pulmonary function is critically important to health and can be assessed by measurement of lung volumes and capacities, *and* by measuring ventilation as a function of time. We will make these measurements in this lab exercise, and observe some effects of exercise on the respiratory system.

An exercising individual has a much higher metabolic rate than one who is quietly relaxing. The demand for oxygen increases tremendously, and the respiratory system keeps pace with this demand by increasing the total minute volume. The total minute volume is the product of the rate and depth of breathing per minute. Total minute volume is adjusted by pulmonary regulatory mechanisms to compensate for changes in metabolic rate. Oxygen consumption per minute is a measure of metabolic rate.

Pulmonary regulatory mechanisms are complex. Chemoreceptors in the medulla and carotid and aortic bodies respond to increases in CO_2 and decreases in pH, and stimulate an increase in ventilation. However, it is believed that other mechanisms are much more relevant in the stimulation of the hyperpnea (i.e. deep breathing which may or may not be accompanied by increased frequency of breaths) of exercise. During exercise there is certainly an increase in CO_2 production but there is also an immediate increase in ventilation. Therefore the blood levels of CO_2 do not rise above normal, and other factors serve as the primary stimulus for the increased ventilation. We will perform a few exercises related to the regulation of ventilation by CO_2 in the lab exercise titled acid base balance.

Objectives

- Define the different lung volumes and capacities, and measure them using spirometry.
- Describe the forced expiratory volume and perform this test.
- Explain how pulmonary function tests are used in diagnosis of restrictive & obstructive disorders.
- Learn how to measure total minute volume and oxygen consumption.
- Describe how these measurements are related, and how they change during and after exercise.

Materials

Collins respirometer
FEV rulers
Functional lung model
Vital capacity tables

Term Definitions

Lung Volumes	The four nonoverlapping components of the total lung capacity.
• Tidal volume	volume of gas inspired or expired in an unforced respiratory cycle
• Inspiratory reserve volume	maximum volume of gas that can be inspired during forced breath in addition to tidal volume
• Expiratory reserve volume	maximum volume of gas that can be expired during forced breath in addition to tidal volume
• Residual volume	volume of gas remaining in the lungs after a maximum expiration
Lung Capacities	Measurements that are the sum of two or more lung volumes
• Total lung capacity	total amount of gas in the lungs after a maximum inspiration
• Vital capacity	maximum amount of gas that can be expired after a maximal inspiration
• Inspiratory capacity	maximum amount of gas that can be inspired after a normal tidal expiration
• Functional residual capacity	amount of gas remaining in the lungs after a normal tidal expiration

Table 13.1–Terms Used to Describe Lung Volumes and Capacities

Procedures

The Collins respirometer will be used for all measurements made in this lab: tidal volume, vital capacity, FEV_1 and oxygen consumption. There are some important general ideas to note about the use of this piece of equipment. When the subject inhales, the bell goes down and the pen goes up; when the subject exhales the bell goes up and the pen goes down. As subject inhales oxygen comes out of the bell; as subject exhales their CO_2 is absorbed by the soda lime crystals in the respirometer. Therefore the only gas in the bell is oxygen. As the experiment progresses there will be less and less oxygen in the bell and we will observe an upward trend of the recordings on the graph paper. The x axis of the of the kymograph chart is in millimeters; the y axis is in milliliters. Because the speed at which the chart moves is known, a graph of air volume (ml) moved into and out of the lungs in a given time interval can be obtained. The temperature and pressure inside the respirometer are different from those existing in the body, therefore the volumes measured must be standardized. To do this the volumes measured are multiplied by a correction factor known as BTPS (body temperature, atmospheric pressure, saturated with water vapor), which is very close to 1.1 at normal room temperatures. Calculate the following data from the prepared kymographs:

Tidal Volume (TV)

Tidal volume is the volume of gas inspired or expired in a normal respiratory cycle. The subject will sit quietly and breathe normally into respirometer. Determine your tidal volume by subtracting the low volume after exhalation from the next high value for inspiration. Repeat for several respiratory cycles and average the results. Remember to then multiply by 1.1, and record this and all other values in lab report.

Inspiratory Reserve Volume (IRV)

IRV is the amount of air you can inhale beyond a normal inspiration. Measure from the top of the last normal inspiration to the top of a maximal inspiration, subtract the low value from the high value, x 1.1.

Expiratory Reserve Volume (ERV)

ERV is the difference between the last normal exhalation and a maximal exhalation, multiplied by 1.1.

Vital Capacity (VC)

VC is the total amount of air that can be expired after a maximal inspiration. Measure the difference between maximal inhalation to maximal exhalation, multiply by 1.1. Next look up the predicted VC on the nomogram. There is one for males, one for females. Note that 2.54 cm / inches will give you height in cm. Calculate your % VC relative to normal: measured VC divided by predicted VC x 100 = % of predicted VC.

Forced Vital Capacity (FVC)

FVC is the total amount of air that can be forcibly expired after a maximal inspiration. In an individual with normal lung function, FVC = VC.

Forced Expiratory Volume (FEV_1)

FEV_1 is ventilation as a function of time, the amount of forced vital capacity (FVC) exhaled in one second. The procedure for obtaining this measurement is that the subject takes a deep, forceful inhalation, holds this momentarily, the kymograph is changed to a fast speed, then the subject exhales as a rapidly and forcefully as possible. At the slow kymograph speed there is one minute between green lines, this distance is one second on the fast speed. After determining FEV_1 from your kymograph, use the equation below to calculate the percent FEV_1:

$$Percent\ FEV_1 = \frac{FEV_1\ (\text{mL})}{FVC\ (\text{mL})} \times 100\%$$

Enter this value in the lab report and refer to Table 13.2 below for predicted values.

Effect of Exercise on Pulmonary Function

In this experiment we will measure the total minute volume and oxygen consumption as a function of exercise. Under resting conditions breathe normally into the respirometer for a few minutes. Determine the frequency of respiration (breaths/minute) and the tidal volume (BTPS corrected). Then have the subject perform light exercise (10 jumping jacks, or 2 minutes on the stationary bike) and then repeat the respirometer measurements. The respiration rate and tidal volume for rest and exercise are read from the graph. Calculate the **total minute volume** by multiplying the frequency of respiration by the tidal volume.

Determine **oxygen consumption** as follows: use a straight edge to draw a line that averages either the peaks or troughs of the tidal volume measurements. Subtract the milliliters where this straight line intersects two vertical chart lines at the beginning and at the end of one minute. The value obtained is the oxygen consumption in ml. Enter values for total minute volume and oxygen consumption measured at rest and after exercise in your lab report.

Age	Predicted Percent FEV_1
18–29	82–80%
30–39	78–77%
40–44	75.5%
45–49	74.5%
50–54	73.5%
55–64	72–70%

Table 13.2 – Predicted Percent FEV_1

Human Physiology Lab 13

Pulmonary Function

Name ______________________________

1. Enter your data for **lung volumes** and **capacities** below:

TV ____________ VC ____________

IRV ____________ predicted VC from nomogram ____________

ERV ____________ your % of predicted VC ____________

Percent $\mathbf{FEV_1}$ = __________ % **Predicted percent** $\mathbf{FEV_1}$ = __________ %

Exercise	Resting	Exercise	% increase
Total minute volume			
Oxygen consumption			

2. Identify the following lung volumes and capacities, give names not volumes:
maximum amount of air that can be inspired after a normal inspiration____________
maximum amount of air that can be inspired after a normal expiration ____________
maximum amount of air that can be expired after a maximum inspiration____________
percentage of forced vital capacity exhaled in the first one second of forced exhalation
____
total minute volume____________

3. Distinguish between obstructive and restrictive pulmonary diseases.

	Physical Problem	Percent FEV_1	VC
Obstructive			
Restrictive			

4. In your third experiment you drew a line that averaged the rising peaks or troughs of tidal volume fluctuations, and then subtracted one y axis value from another that was one minute apart. Why does this value give you a measure of oxygen consumption?

5. Define and describe the following respiratory conditions:

asthma ____________
atelectasis ____________
bronchitis ____________
emphysema ____________
pneumothorax ____________
pneumonia ____________
tuberculosis ____________
pulmonary fibrosis ____________
respiratory distress syndrome ____________

Laboratory Exercise 14

Renal Function

Introduction

The kidneys play a major role in excretion and osmoregulation. The kidneys are responsible for the elimination of most of the waste products of metabolism as well as any unusual substances or drugs we may have taken into our bodies. In their regulatory role, the kidneys are involved in the homeostatic regulation of fluid balance, electrolyte balance and acid–base balance. The kidneys perform these functions primarily by producing urine. Blood is first driven into nephrons by hydrostatic pressure, producing an ultrafiltrate of blood. As this filtrate passes through the renal tubules, 99% of this fluid is reabsorbed, maintaining fluid volume. Selective reabsorption of ions, Na^+, K^+, Cl^-, H^+, HCO_3^-, maintains electrolyte and acid base balance. Essential metabolites such as sugars and amino acids are also reabsorbed.

Under pathological conditions, substances may appear in the urine which are diagnostic of metabolic disorders. One can also find traces of any drugs that have been ingested or injected. Thus, examination of urine (urinalysis) is of great clinical importance. In this lab, you will perform several parts of a standard urinalysis, including direct observation, tests of chemical composition, and microscopic examination of urine sediment. You will also challenge your kidneys with a water or a salt load and observe the compensatory increases in the urinary excretion of these substances.

Because you will be observing the effect of a water or salt load on kidney function, you must stop eating, drinking and urinating one hour before lab starts. Therefore, be sure to eat a good breakfast or lunch before your lab time.

Objectives

- Observe normal constituents of urine & urine sediment.
- Learn what pathological conditions are indicated by changes in the constituents of urine.
- Explain how the kidneys respond to water and salt loading.
- Learn how to measure urine volume and electrolyte composition.
- On models identify the major anatomical structures of the kidney: nephron: renal corpuscle: Bowman's capsule + glomerulus, PCT, loop of Henle, DCT; collecting duct; juxtaglomerular apparatus: JG cells + macula densa; afferent & efferent arterioles; peritubular capillaries; vasa recta

Materials

Urine collection cups
Microscopes, slides and coverslips
Centrifuge and test tubes
Multistix
Potassium chromate solution
Silver nitrate solution
Kidney models
Urinalysis reference texts

Procedures

The procedures for the urinalysis and the water / salt load experiment will be described separately. Begin the water/salt experiment first, as several urine collections are required. You can perform the urinalysis while waiting for the next urine collection time. For the water / salt experiment you need to know the volume of urine, so you collect all of it, rather than the usual midstream sample collected in clinical labs.

Renal Regulation of Fluid and Electrolyte Balance (Experiment #1)

In this experiment, you will void your urine multiple times, each time recording the total volume voided and each time returning with a urine sample. The urine sample will be used to complete a variety of assays. For each sample, both the volume voided and the chloride concentration will be recorded in Table 14.1 on the next page.

1. Students must void their urine at the beginning of the laboratory session. This will be the time zero sample and will be used for urinalysis.

2. Every student drinks 500 ml water, and a special group of volunteers ingests salt tablets in addition to the water. Students then void their urine every 30 minutes for 2 hours. The urine sample volumes are measured each time, and are tested for chloride concentration. All individual data are recorded in Table 14.1 and on the class data sheet. There will be 5 samples taken, at 0, 30, 60, 90 & 120 minutes.

3. Chloride concentration is determined by a simple chemical test.
   Measure 10 drops of urine into a test tube.
   Add 1 drop of 20% potassium chromate solution.
   Add 2.9% silver nitrate solution, one drop at a time, while shaking the tube continuously.
   Count the minimum # of drops that change the color of the solution from yellow to brown and record the number of drops in Table 14.1.

   Note that we measure chlorine rather than sodium because there is a simple color assay for chlorine. The urine sample contains chlorine, and silver will preferentially bind to chlorine. As you add silver (silver nitrate) it binds first to chlorine, and then when all the chlorine has been bound, the silver will bind to chromate, turning the solution brown.

4. Calculation of milliequivalents of Chloride.
   The concentration of ions in body fluids are usually given in terms of milliequivalents (mEq) per liter (mEq/L). The advantage of this is that the total concentration of anions can easily be compared with the total concentration of cations. Usually these values are each 156 mEq/L. To conform to this standard we need to express our data in mEq/L.

   Convert the number of drops of silver nitrate to mEq/L of chloride using the equation below. The concentration of sodium will be the same since sodium and chloride exist in a 1 to 1 ratio. Record your data in Table 14.1. For a complete explanation of this equation see the Notes at the end of this lab. (*Round your calculations to whole numbers.*)

   # drops x 61 x 10 divided by 35.5 g = amount Cl in your sample in mEq/L

Renal Data				
Time (min)	Urine Volume (mL)	# Drops of Silver Nitrate Added	Cl^+ Concentration (mEq/L)	Na^+ Concentration (mEq/L)
0				
30				
60				
90				
120				

Table 14.1 – Renal lab data

Urinalysis (Experiemnt #2)

1. **Direct observation**: note the color of your urine and review Table 14.2 for the clinical correlate of various colors.
2. The **chemical tests** are all performed with test strips called Multistix. Dip one of these in your first (time 0) urine sample. Place the Multistiz on its side on the napkin to prevent cross contamination of chemical pads. Wait the specified time indicated on the Multistix bottle, and compare the colors of all the little square chemical pads with the color chart on the Multistix bottle. Each chemical pad on the Multistix tests for the presence of one of the following: leukocytes, nitrite, urobilinogen, protein, pH, blood, specific gravity, ketones, bilirubin and glucose. Because virtually all students will have normal urine, there will be three beakers which have had different substances added to the urine. Test these as well as your own sample, and record what you think has been added to each of the 3 "mystery urine samples" in your lab report. You will need to refer to your text, and reference books in the lab to learn the clinical correlates for abnormal values for these chemicals.
3. **Microscopic evaluation** is conducted on stained urine sediment. Fill the plastic centrifuge tube (blue cap) to 10mL and give to the lab instructor to be centrifuged. The lab instructor will wait until the centrifuge is full and return your sample to you after it has been spun. Carefully discard the supernatant by pipetting out this fluid without disrupting the sediment pellet at the bottom of the tube. If there is no visible pellet, carefully pipette out the fluid leaving just a drop or two of undisturbed fluid at the bottom of the tube. Discard the supernatent into your urine cup, and place a drop of stain on the sediment or remaining fluid. Mix these together and pipette a drop for observation. Place it on a microscope slide, cover with a coverslip, and identify what you see. The urinalysis reference texts in the lab will help with identification, there is also information on the next page for reference. You will again need reference texts to obtain information on the clinical correlates for the crystals, cells, bacteria and casts which may be seen.

Color	Cause
Yellow–orange to brown–green	bilirubin from obstructive jaundice
Red, red–brown, smokey red	unhemolyzed RBCs from urinary tract
Dark wine	hemolytic jaundice
Brown–black	melanin pigment from melanoma
Dark brown	hepatitis, pernicious anemia, malaria
Green	bacterial infection (Psuedomonas aeruginosa)

Table 14.2 – Appearance of Urine and Cause

Interpretation of Urinalysis Data – Explanation of chemical test results (Multistix test strips)

Leukocyte (LEU) – a small amount of leukocytes in the urine is normal. Elevated leukocytes in urine indicates a urinary tract infection (UTI) that can occur in any part of the urinary system including kidneys, ureters, bladder and urethra. A positive leukocyte test with a positive nitrites test is normally indicative of infection.

Nitrite (NIT) – Metabolism of proteins involves deamination and urea is created from the amino groups. Nitrites should not be present as a normal metabolic product. However, they may be present in the case of *E coli* infections, as they are waste products of these bacteria.

Urobilinogen (URO) – Bile is metabolized in the small intestine by bacteria, and urobilinogen is a normal breakdown product of bile. Any time an excess of RBCs are destroyed suddenly, urobilinogen can appear in the urine. A small amount is normal, an excess may indicate anemia, polycythemia or a transfusion reaction.

Protein (PRO) – Proteins are usually too large to enter the glomerular filtrate. Proteins in urine may indicate that capillaries have become more permeable, perhaps due to glomerulonephritis (renal infection).

pH – The normal pH range for urine is 5–7. Values differing from this range may be due to unusual nutritional habits, metabolic abnormalities, or be the result of medications someone is taking.

Blood (BLO) – RBCs do not generally enter the glomerular filtrate. Blood may appear as whole cells or as hemoglobin from lysed cells. Blood or hemoglobin in the urine may indicate kidney stones, tumors, urinary tract infections, hemorrhage, clotting disorders, or the rupture of glomerular capillaries. Finding some blood in the urine of menstruating women is normal. A few white blood cells can be normal, many indicate an inflamatory process, in kidney or urinary tract.

Specific Gravity (SG) – is the ratio of the density of a substance to the density of water. Normal urine specific gravity can range from 1.001-1.030. The specific gravity of urine will increase with dehydration and will be abnormally high in disease states such as diabetes insipidus where the absence of antidiuretic hormone causes great loss of water by the kidneys.

Ketone (KET) – If the body cannot utilize glucose for its energy needs, it utilizes fatty acids as an alternative energy source. Intermediates of fatty acid metabolism are called ketone bodies. If the body is relying heavily on fatty acids for its energy needs, the ketone bodies will be excreted in the urine. This is another indication of diabetes mellitus, as well as a low carbohydrate diet.

Bilirubin (BIL) – Old RBCs are destroyed in the spleen and liver. The heme groups of hemoglobin are turned into bilirubin that is sent to the liver where it is turned into bile and secreted into the GI tract. Elevated bilirubin levels in the urine may indicate an increased RBC destruction or the inability of the liver to process bilirubin due to hepatitis or cirrhosis. A blocked bile duct may result in bilirubinuria.

Glucose (GLU) – Glucose should pass through the glomeruli easily but be completely reabsorbed, if glucose levels are within normal homeostatic ranges. If glucose appears in the urine, it usually indicates that blood glucose levels are above homeostatic ranges and are overloading the glucose transporters in the proximal convoluted tubules. This usually occurs in diabetes mellitus. *Note that the appearance of glucose in the urine of a diabetic does *not* represent a regulatory mechanism for the removal of excess glucose from the blood!*

Microscopic evaluation of urine

Crystals – can form from many different chemicals. They usually have no clinical significance, but can relate to formation of urinary calculi (stones).

Casts – are cylindrical structures formed by the precipitation of proteins in the renal tubules. They may form if there is a marked decrease in urine flow, and may indicate intrinsic renal disease.

Human Physiology Lab 14 Name ____________________________

Renal Function

1. Clinical examination of urine, using Multistix and microscopic examination.

Substance	**Your Data**	**Clinical Significance** *assume test is positive in each case*
Leukocyte		
Nitrite		
Urobilinogen		
Protein		
pH		
Blood		
Specific Gravity		
Ketone		
Bilirubin		
Glucose		
Casts		
Crystals		

2. Mystery urine – what unusual results were found with multistix?

   A ____________________ B ____________________ C ____________________

3. Describe in words what is meant by the term "renal plasma threshold for glucose"

4. Review the graph of class data and be prepared to write a conclusion from these data, commenting on the mechanisms which account for changes seen in the volume and salt concentration of the urine.

   What hormone regulates plasma sodium concentration? ________________

   What hormone regulates plasma osmolarity or water content? ________________

   What are your conclusions? Include mechanisms.

5. Using your text, name the different regions of the nephron affected by the following diuretics:

   Loop diuretics ______________________________

   Thiazide diuretics ______________________________

   Osmotic diuretics ______________________________

   Carbonic anhydrase inhibitors ______________________________

Notes:

First, multiply the number of drops of silver nitrate added by 61 to obtain chloride concentration in mg/100ml. Because each drop of 2.9% silver nitrate that you add to your urine sample is equivalent to 61 mg Cl^-/100 ml urine, multiply the number of drops added by 61 to obtain [Cl^-] as mg/dl.

Next, we must convert mg/dl to millimoles/L. First, multiply by 10 to get the value in L vs. dl. The atomic weight of Cl^- is 35.5 g, which means that 1 mole contains 35.5 g. Remember ratios from Lab Exercise 1!

$$\frac{1 \text{ mole}}{35.5 \text{ g}} = \frac{? \text{ moles}}{\# \text{ g you found in 1 L}} \qquad \left(\text{divide \# g/L by 35.5 g}\right)$$

Laboratory Exercise 15

Acid Base Balance

Introduction

It is critical for the body's homeostasis that the concentration of **hydrogen ions** in the blood be maintained within the narrow range of **pH** 7.35 to 7.45. Death occurs below 7.0 and above 7.8; most enzymes can't operate properly if the pH is outside this range and neural activity is compromised. It is therefore important to learn what pH is and how it is regulated in the body. The regulation of pH is accomplished by **buffers**, and the respiratory and renal systems. In this lab you will observe the power of a buffer to resist changes in pH, and the role of the respiratory system in acid–base balance. You will also integrate your knowledge of pH regulation by attempting to analyze the acid base balance of some hypothetical patients.

Ventilation has two different functions, oxygenation of blood and elimination of carbon dioxide. The latter serves to maintain the normal pH of the blood. Regulation of ventilation is primarily geared to levels of CO_2 and H^+ in the blood. If levels of CO_2 rise, levels of H^+ will also rise and ventilation will increase. Conversely, if CO_2 levels fall, ventilation will decrease. During exercise, regulation of ventilation is more complex. Ventilation increases before CO_2 levels have risen significantly. The increased rate of CO_2 production during exercise is matched by an increase in the rate of its elimination through ventilation. This increase in CO_2 production after exercise will be demonstrated in a very simple experiment.

Use your text to review the meaning of the terms: **pH, acid, base, buffer, acidosis, alkalosis**. Acidosis and alkalosis can occur for a variety of reasons, linked to either metabolic or respiratory problems. If the problem is metabolic, both respiratory and renal systems can compensate. If the problem is respiratory, only the renal system can compensate. In this lab we will solve some acid base problems of hypothetical patients to gain experience with these interrelationships.

Objectives

- Observe the ability of a buffer to stabilize the pH of solutions.
- Explain the relationships between different respiration rates, CO_2 production, pH.
- Observe the effect of exercise on CO_2 production.
- Practice using blood values for pH, PCO_2, HCO_3^- to determine the cause of acidosis or alkalosis.

Materials

pH meter, phosphate buffer, 2N NaOH, 2N HCl
Beakers with water, phenolphthalein, straws

Procedures

Buffers

1. Allow the pH meter to warm up in the standby position in its buffer solution. Be sure the electrode is always immersed in fluid and not allowed to dry.
2. Before placing the pH meter into a new solution or returning it to its original buffer solution, it should first be rinsed into a waste beaker using deionized water and then blotted dry with a Kimwipe.
3. Immerse the pH electrode in 150 ml deionized H_2O in a 250 ml beaker and record the pH.
4. Add 1 drop of concentrated HCl, mix, record the pH. Add 2 additional drops of HCl, mix, and record the new pH. Be careful not to add the acid directly on the electrode.
5. Repeat this procedure, starting with fresh de–ionized water and adding first 1 and then 2 additional drops of concentrated NaOH, recording the pH after one and three drops.
6. Repeat this procedure starting with phosphate buffer instead of water. Read the pH, add 1+ 2 drops HCl, and then 1+2 drops NaOH, using fresh buffer.

Exercise and CO_2 Production

1. Add ~20 ml phenol red dilute solution to a beaker. The phenolphthalein is a pH indicator which turns pink in alkaline solutions and clear or orangy colored in neutral or acidic solutions.
2. Sit quietly and exhale through a straw into the solution in the first beaker. Note the time it takes to turn the solution from pink to orange. Time ____________. Empty and rinse beaker.
3. Add another 20 ml phenol red dilute solution to your beaker.
4. Exercise vigorously for 5 minutes. Exhale through a straw into the solution, and again note the time it takes to clear the pink solution. Time ____________. Empty and rinse beaker.

Acidosis and Alkalosis

1. Read the exercise taken from *Human Anatomy and Physiology* by E. N. Marieb provided on the following page, and solve the problems. ***Note***: a problem similar to those on the next page will be included in your next lab exam!

Using Blood Values to Determine the Cause of Acidosis or Alkalosis

Students, particularly nursing students, are often provided with blood values and asked to determine (1) whether the patient is in acidosis or alkalosis, (2) the cause of the condition (respiratory or metabolic), and (3) whether or not the condition is being compensated. Such determinations are not nearly a difficult as they may appear if they are approached systematically. When attempting to analyze a person's acid–base balance, scrutinize the blood values in the following order:

1. Note the pH. This tells you whether the person is in acidosis (pH < 7.35) or alkalosis (pH > 7.45), but it does *not* tell you the cause.
2. Next, check the PCO_2 to see if this is the cause of the acid–base imbalance. Because the respiratory system is a fast–acting system, an excessively high or low PCO_2 may indicate either that the condition is respiratory system–caused or compensating. For example, if the pH indicates acidosis and (a) PCO_2 is over 45 mm Hg, the respiratory system *is the cause* of the problem and the condition is a respiratory acidosis; (b) PCO_2 is below the normal limits (below 35 mm Hg), the respiratory system is *not the cause but is compensating;* (c) PCO_2 is within normal limits, the condition is *neither caused nor compensated* by the respiratory system.
3. Check the bicarbonate level. If step 2 proves that the respiratory system is not responsible for the imbalance, then the condition is metabolic and should be reflected in increased or decreased bicarbonate levels: Metabolic acidosis is indicated by HCO_3^- values below 22 mEq/L, and metablic alkalosis by values over 26 mEq/L. Notice that whereas PCO_2 levels vary inversely with blood pH (PCO_2 rises as blood pH falls), HCO_3^- levels vary directly with blood pH (increased HCO_3^- results in increased pH).

Consider two examples of this approach:

Problem 1

Blood values given: pH – 7.5; PCO_2 – 24mm Hg; HCO_3^- 24 mEq/L.

Analysis:

1. The pH is elevated = alkalosis.
2. The PCO_2 is very low = the cause of the alkalosis.
3. The HCO_3^- value is within normal limits.

Conclusion: This is a respiratory alkalosis not compensated by renal mechanisms, as might occur during short–term hyperventilation.

Problem 2

Blood values given: pH – 7.48; PCO_2 – 46 mm Hg; HCO_3^- 32 mEq/L.

Analysis:

1. The pH is elevated = alkalosis.
2. The PCO_2 is is elevated = the cause of *acidosis*, not alkalosis; thus, the respiratory system is compensating and is not the cause
3. The HCO_3^- is elevated = the cause of the alkalosis.

Conclusion: This is a metabolic alkalosis being compensated by respiratory acidosis (retention of CO_2 to restore blood pH to the normal range).

A simple chart to help you in your future determinations is provided below.

Normal range in plasma	pH 7.35–7.45	PCO_2 35–45 mmHg	HCO_3^- 22–26mEq/L
Acid–base disturbance			
Respiratory acidosis	↓	↑	↑ if compensating
Respiratory alkalosis	↑	↓	↓ if compensating
Metabolic acidosis	↓	↓ if compensating	↓
Metabolic alkalosis	↑	↑ if compensating	↑

Human Physiology Lab 15

Acid Base Balance

Name ________________________________

1. Buffers

pH of distilled water	__________	pH of buffer	__________
pH of water + 1 drop HCl	__________	pH of buffer + 1 drop HCl	__________
pH of water + 3 drops HCl	__________	pH of buffer + 3 drops HCl	__________
pH of water + 1 drop NaOH	__________	pH of buffer + 1 drop NaOH	__________
pH of water + 3 drops NaOH	__________	pH of buffer + 3 drops NaOH	__________

Do your data support the statement that buffers help to stabilize the pH of solutions? Explain.

2. Exercise and CO_2 Production

Time for color change at rest: __________ Time for color change after exercise: __________

What is the effect of exercise on CO_2 production? How did your results demonstrate this effect?

3. Differentiate hyperventilation and hyperpnea.

4. State whether the following cause an increase or a decrease in respiratory rate and depth:

increased blood pCO_2	________________	decreased blood pH	________________
decreased blood pO_2	________________	increased blood pH	________________

5. Acidosis and Alkalosis

A patient is in acidosis when the pH is below______; in alkalosis when the pH is above ______

If a patient is in alkalosis, and the PCO_2 is elevated, is the respiratory system the cause or is it compensating? ____________________________ Explain your answer.

Laboratory Exercise 16

Digestion

Introduction

The digestive system provides nutrients to the body. These are ingested and must be absorbed (transported through the epithelial membrane lining the digestive tract) to reach the blood for distribution to the cells of the body. Foods must be broken down by physical processes (chewing, stomach churning) and enzymatic hydrolysis before nutrients can be absorbed. The process of hydrolyzing larger food molecules (polymers or macromolecules) into absorbable monomers is known as **digestion**.

The digestion of each of the major polymers, carbohydrate, protein and fat, is accomplished by the action of **specific** enzymes. The fact that an enzyme is specific means it will catalyze only one reaction; only one ligand will bind to the functional binding site. Many organs of the digestive system synthesize and secrete these enzymes into the lumen of the tract. The major digestive enzymes are listed in Table 16.1. In this lab you will observe the effects of some of these enzymes in test tubes. Note that these enzymes work extracellularly in your body and thus should also work readily in test tube environment, given the correct solvent, concentration, pH and temperature. You will add substrate (polymer) and enzyme to a test tube, and replicate the temperature and pH conditions under which these enzymes are most active. You will include **control** tubes in your experiments, to make sure that the action observed is due to the enzyme and not some other factor. For each enzyme tested, there must be some way to identify whether a hydrolytic reaction has occurred. There are two possible approaches to this problem. You can measure the disappearance of **substrate** (polymer), or the appearance of **product** (monomer). You will use both approaches in this lab.

Objectives

- Know which enzymes digest which macromolecules.
- Explain the meaning of the term specificity, and apply it to enzyme action.
- Recognize factors which may be necessary or inhibitory for enzymatic action, in lab and GI tract.
- Describe the significance of a control in experiments.
- Distinguish enzymatic digestion and emulsification.
- Observe the effects of some of the digestive enzymes.
- Learn assay procedures for some polymers and monomers.

Materials

Water and ice baths
Starch solution, Benedict's reagent, Lugol's solution (iodine), amylase, maltose
Protein (collagen on photographic film), pepsin, HCl and NaOH solutions

Procedures

Details about the procedures will be given by the lab instructor.
One basic experimental design is presented here, details may vary each semester.

Digestion of Carbohydrate by Amylase

The carbohydrate used will be starch.
The enzyme is salivary amylase (students donate the enzyme!).
There are assays for both polymer (starch) and monomer (the disaccharide maltose).
Starch test: add iodine (*Lugol's reagent*).

A positive test for the presence of starch is indicated by a purplish black color.

Maltose test: the hydrolysis of starch produces the product maltose, a reducing sugar. *Benedict's test* indicates the presence of reducing sugars. Benedict's reagent contains cupric ions (Cu^{2+}) which, in the presence of maltose, are reduced to cuprous ions (Cu^{+}) that form a yellow colored precipitate (Cu_2O). Results of the Benedict's test are evaluated as follows:

Blue (no maltose)	–
Green	+
Yellow	++
Orange	+++
Red (most maltose)	++++

Digestion of Protein by Pepsin

The protein used will be a component of photographic film, usually made from collagen protein or gelatin. The collagen protein is used to bind the photographic crystals to the film strip. The enzyme being tested is pepsin, a gastric enzyme with a pH optimum adapted to the normal pH of the stomach (pH 1-2). The assay is simply the observation of a change in a small square of film, from dark to clear. Dark film indicates no protein digestion. Clear film indicates complete protein digestion. Opaque film means that some collagen protein remains on the film which indicates partial protein digestion.

Organ	Substrate	Enzyme	Products of Enzyme Action	Optimal pH
Mouth	Carbohydrate			
	Starch	Salivary amylase	Disaccharides	6.7
Stomach	Protein	Pepsin	Peptides	1.6–2.4
Pancreas	Carbohydrate			
	Starch	Pancreatic amylase	Disaccharides	6.7–7.0
	Protein	Trypsin	Polypeptides	8.0
		Chymotrypsin	Dipeptides	8.0
		Carboxypeptidase	Amino acids	8.0
	Lipid	Lipase	Glycerol and fatty acids	8.0
Small Intestine	Carbohydrate			
	Maltose	Maltase	Glucose	5.0–7.0
	Lactose	Lactase	Glucose and galactose	5.0–7.0
	Sucrose	Sucrase	Glucose and fructose	5.0–7.0
	Protein	Aminopeptidase	Amino acids	8.0
		Carboxypeptidase	Amino acids	8.0

Table 16.1 – Selected enzymes of the digestive system

Carbohydrate Digestion
Procedure

1. Each group will add varying reagents, as described below, to a single test tube.
2. Mix the contents of each tube.
3. Test tubes will be incubated for 1 hour in a 37 °C water bath.
4. After incubation mix contents of each tube & divide contents in half by pouring half into a clean tube.
   a) Test one tube for starch by adding three drops of iodine solution (Lugol's reagent in a brown bottle). A positive test is indicated by the development of a purplish/black color.
   b) Test the other tube for reducing sugars (maltose) as follows:
      add 1 pump (5.0 mL) Benedict's reagent to the test tube
      immerse tube in boiling water bath for 2 minutes
      remove tube and assess amount of reducing sugar by the color scale on previous page
         [if directed to, use prepared solutions of glucose and sucrose as standards]

Test Tube #	add 3.0 ml	add 5.0 ml	other reagents
1	water	starch solution	none
2	saliva	starch solution	none
3	pepsin	starch solution	none
4	saliva, boiled	starch solution	none
5*	saliva	starch solution	10 drops conc. HCl

*Group 5: *add saliva to test tube first, then 2 N HCl, then add the starch, and shake well*

Record results below and on the class data sheet provided for class discussion.

Test Tube #	starch present?	maltose rating	conclusion
1			
2			
3			
4			
5			

Protein Digestion
Procedure

There are 8 test tubes with 2 black marks each. Add the following to the appropriate tube: first solution to the first black mark (2cm) and second solution to the second black mark (4cm) (solutions are: 1.5% pepsin; 0.8% HCl; 0.1% NaOH)

1. pepsin solution to first mark; HCl to second mark
2. boiled pepsin solution to first mark; HCl to second mark
3. pepsin solution to first mark; NaOH to second mark
4. pepsin solution to first mark; deionized water to second mark
5. pepsin solution to first mark; HCl to second mark
6. deionized water to first mark; HCl to second mark
7. saliva to first mark; HCl to second mark
8. pepsin solution to first mark; HCl to second mark; 0 °C

Mix the contents of each tube.
Measure the pH of each tube using pH paper, record below
Add a 1x1 cm piece of film to each tube, shake to move film to bottom of tube.
Place tubes # 1–7 in 37 °C water bath; tube # 8 in a 0 °C ice bath.
Incubate the tubes for one hour.
Every 20 minutes agitate tubes and check color of film.
After the incubation period, remove film and place on paper towel under appropriate number.
A positive result (digestion of protein on film) will be indicated by film changing from dark to clear.

Record results below and on the class data sheet.
Describe the **Amount of Film Coating** as: none, some.
Assign a rating for **Pepsin Activty** as: – , + , + + +

Test Tube	Contents	pH	Amount of Film Coating	Pepsin Activity
1	pepsin, HCl			
2	boiled pepsin, HCl			
3	pepsin, NaOH			
4	pepsin, water			
5	pepsin, HCl			
6	water, HCl			
7	amylase, HCl			
8	pepsin, HCl, cold			

Human Physiology Lab 16 Name ____________________

Digestion

1. Fill in this table, which outlines the molecules and tests (assays) used in this experiment.

polymer	enzyme	substrate	substrate test	product	product test
carbohydrate					
protein					N/A

2. In the space provided, explain each of these aspects of the experimental design:

Experimental protocol	Explanation
37 °C incubation period	
HCl	
NaOH	
boiled enzyme	
omission of enzyme	
different enzyme	

3. Differentiate the following terms:
   Digestion and emulsification

   Chylomicrons, HDLs, and LDLs

4. What is meant by the term specificity?

5. What is the purpose of a control in an experiment?

6. Be sure you can describe the results for *both* the carbohydrate and protein experiments. Did your results confirm what you expected about enzyme activity? Enzyme specificity?

Laboratory Exercise 17

Glucose Tolerance

Introduction

One of the homeostatic functions of the body is to regulate blood fuel levels. Sugars, amino acids, fatty acids, and glycerol are the principal sources of energy for cellular metabolism. The preferred energy molecule by many cells is the monosaccharide glucose, and its fasting blood level is tightly regulated within a normal range of 70–99 mg/dl (this value varies by demographics). Cells in the islets of Langerhans in the pancreas monitor blood glucose levels. A decrease in plasma concentration of glucose stimulates the alpha (α) cells to release glucagon, which elevates plasma glucose levels by stimulating glycogenolysis and glugoneogenesis in the liver. An increase in the plasma concentration of glucose stimulates beta (β) cells to release insulin, which lowers plasma glucose levels by stimulating glycogenesis in liver and skeletal muscle and by increasing cellular uptake of glucose.

Diabetes mellitus is a disease where the body either does not produce or does not respond adequately to insulin resulting in abnormally high blood glucose levels. In type I diabetes mellitus, the pancreas secretes little to no insulin in response to a glucose load. In type II diabetes mellitus, the tissues no longer respond adequately to circulating insulin. One of the diagnostic tests for type II diabetes is the glucose tolerance test. This test examines how well the body responds to a glucose load. Glucose is consumed after an overnight fast and blood glucose levels are determined at 30-minute intervals over the following two hours.

Because the glucose tolerance test determines how a fasting individual responds to a glucose load, it can be used to help diagnose diabetes type I and II. However, in this lab because we are not trying to diagnose, the test will not be performed on individuals known to have type I diabetes. Rather, we will observe the homeostatic regulation of blood sugar levels in normal fasted individuals. We expect that after ingesting glucose, fasting volunteers will show a significant increase in blood glucose followed by a return to baseline, demonstrating the regulatory homeostatic action of insulin.

Objectives

- Perform a glucose tolerance test.
- Describe the role of the pancreas in regulation of blood glucose levels.

Materials

Glucose test strips or glucometer
Glucose solution (75g glucose)
Lancets and alcohol swabs

Procedures

- Volunteers have self-selected for this experiment.
- They must fast for 12-16 hours before the experiment to ensure baseline glucose levels. Fasting means ingestion of nothing except water.
- The fasting volunteers will first obtain a time 0 blood glucose value. A drop of blood will be obtained using a fingerstick procedure and glucose values determined.
- Immediately after the first finger-stick is performed, the subjects will drink a concentrated glucose solution. This drink contains 75 grams of glucose in 296 mls carbonated water. The entire bottle must be drunk. After drinking this solution no other food or beverage, including water, may be ingested until the end of the experiment.
- Blood glucose will be measured again at 30, 60, 90, and 120 minutes.

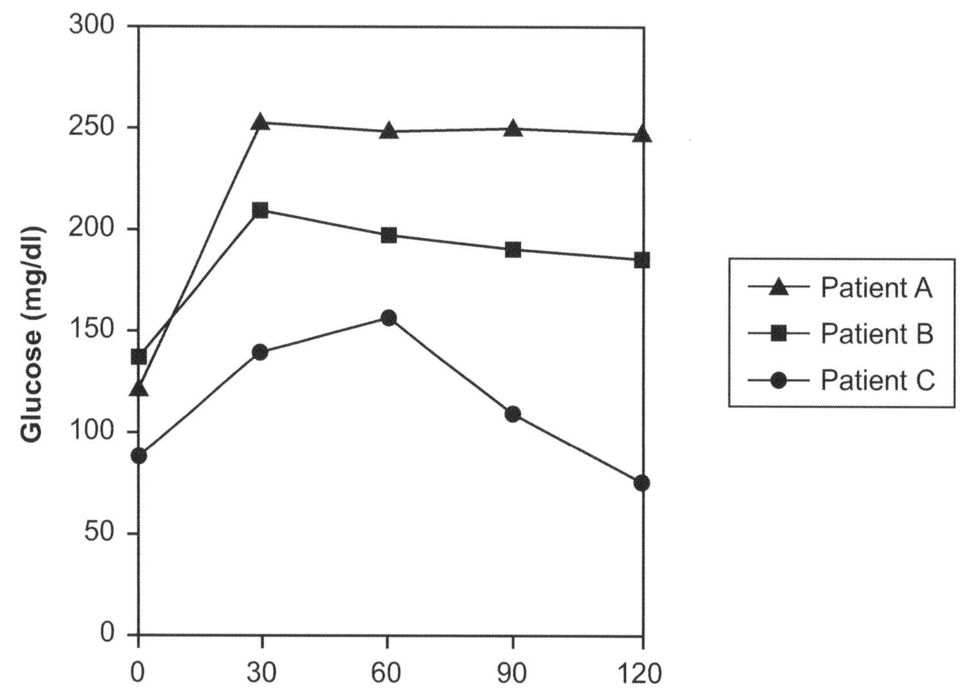

Figure 17.1–Glucose Tolerance Test. Patient C is normal, patient B suffers from type II diabetes, patient A suffers from type I diabetes

Fasting Blood Glucose	
70-99 mg/dL	normal
100-125 mg/dL	pre-diabetic
126 mg/dL & above	diabetes

Glucose Tolerance Test 2 hours after glucose drink	
less than 140 mg/dL	normal
140-200 mg/dL	pre-diabetic
Over 200 mg/dL	diabetes

Table 17.2 Glucose Tolerance Test Values

Human Physiology Lab 17

Name ____________________________________

Glucose Tolerance

1. What are two purposes of a glucose tolerance test?

2. The data from all subjects will be graphed by your instructor. Write a brief conclusion about the results you observed.

3. What are some possible sources of error in this experiment? Name at least two.

4. Differentiate diabetes mellitus and diabetes insipidus.

5. Define these terms:

   glycogenesis ____________________________________

   gluconeogenesis ____________________________________

   glycogenolysis ____________________________________

   ketone bodies ____________________________________

   ketoacidosis ____________________________________

Blood

Introduction

Blood is a connective tissue, consisting of cells (often referred to as formed elements) and a fluid matrix called plasma. The various components of blood serve different physiological functions. The plasma transports nutrients to cells, wastes to the kidneys, and hormones to their target cells. The oxygen carrying function of the blood is performed by hemoglobin, contained in **RBC**s. Measurements of the oxygen carrying capacity of the blood include a red blood cell count, hemoglobin concentration, and hematocrit. When one or more of these measurements is low (see Table 20.1 for normal values), a person may have **anemia**. You will perform these three tests, as well as calculating values (MCV and MCHC) that allow clinicians to diagnose one specific type of anemia vs. another. **Platelets**, along with certain plasma proteins, are involved in the clotting process. The white blood cells (**WBC**s, leukocytes) are involved in immune function. There are five kinds of leukocytes, each has a different immune function. A **differential WBC count** is done to determine the percentage of each class of leukocyte, and can be useful in the diagnosis of many conditions.

Another important blood test is the determination of **blood type**. Blood transfusions require compatibility between donor and recipient blood. Incompatibilities arise due to the presence of characteristic molecules on the surface of red blood cell membranes that can be different in different people. These molecules can function as antigens, and one person's antigens can bind to specific antibodies that may be present in the plasma of a person with a different blood type. The major blood group antigens are the **ABO** system and the **Rh** antigen. Although determining blood type is primarily of clinical importance, it also provides a laboratory view of one way that antibodies can attack and neutralize antigens. This process, called **agglutination**, is the result of **antibody** molecules cross–linking red blood cells that share a particular antigen. Blood type is determined by adding various antisera (antibodies) to blood and looking for the binding of RBC antigens with antisera antibodies, which leads to cross linking and therefore clumping of RBC's (agglutination).

Objectives

- Become familiar with and practice universal precautions.
- Perform a red blood cell count.
- Make hematocrit and hemoglobin measurements.
- Calculate red cell indices that can be used to identify various types of anemia.
- Be able to identify RBCs, neutrophils, lymphocytes, and platelets and know their functions.
- Explain what is meant by the term blood group, determine which blood group you belong to.
- Explain how agglutination occurs, differentiate agglutination and coagulation.
- Know which blood types are compatible for transfusion.

Materials

Lancet, alcohol swabs
Microscope
Hemocytometer, coverslips
Microtube with diluting solution
Stained blood smear slides
Micropipette, 5 ul
Tallquist papers
Capillary tubes and centrifuge
Anti– A, B, and Rh sera
Blood typing tray; toothpicks

Procedures

Universal Precautions

You will be given explicit directions by your lab instructor concerning universal precautions. Universal precautions are actions you take to prevent the transmission of disease from potentially infected body fluids. The universal precautions you will be asked to take in this lab include:

- Work only with your own blood.
- Wear gloves at all times.
- Place sharp objects (broken lancets, toothpicks, coverslips, micropipette tips) in Sharp's Container after contact with blood.
- Dispose of all other materials that contact blood in biohazard bag.
- At end of experiment wash lab bench with disinfectant, hands with soap and water.

NOTE: you will make 4 different observations on blood. Listen to directions carefully, and before you collect blood have all necessary equipment organized so that you only have to prick your finger ONE time. Collect blood in the following order:

- capillary tube
- microtube
- blood typing tray (3 wells)
- Tallquist paper

Complete the following in the order given. All of your data is recorded in the table on the last page of this lab.

1) Hematocrit

Fill a capillary tube with blood by capillary action. Fill from the end that is marked with a red line. To prevent air bubbles, ALWAYS keep the capillary tube horizontal. Using the small plugs, seal the end from which you filled the tube with blood, and place this plugged end facing outward in the centrifuge. After centrifugation, use the hematocrit guide to determine your hematocrit value, and enter this value in your lab report.

2) Blood Type

- Add one drop of fingertip blood to each well.
- Then add one drop of each antisera, A, B, and Rh, in the labeled wells. Mix each well with a different toothpick.
- Examine the wells for agglutination, which indicates a positive reaction. The mixture will look grainy if agglutination has occurred. Record your data in your lab report and on the class data sheet.

3) RBC count

The lab instructor will give directions for collecting blood in the microtube, which contains a premeasured diluting solution. A special slide called a **hemocytometer** is used for making the RBC count. A picture of the hemocytometer grid is shown in Figure 18. Place a coverslip on the hemocytometer. Place two drops of diluted blood from the microtube at the edge of the coverslip. It will be drawn underneath by capillary action. Count the total number of RBCs in five of the small squares in the center of the hemocytometer. They are numbered 1-5 in Figure 18 and include the 4 squares at the corners and the most central square. If a cell is on the left or lower line, include it in your count. Exclude those on the right or upper lines. The central grid of 25 squares is 1 mm^2 in area and 0.10 mm deep. The dilution factor is 1:200. To convert the number of RBCs that you counted in 5 squares to the number per cubic millimeter (mm^3), multiply your count by 10,000 (5x10x200).

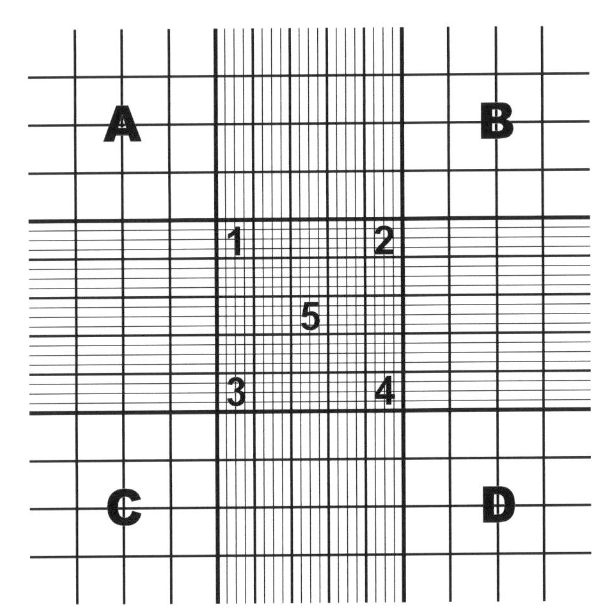

Figure 18–The hemocytometer grid. Squares 1–5 are used for RBC counts; squares A–D are used for WBC counts

4) Hemoglobin
The lab instructor will give directions for determining the hemoglobin concentration of a sample of your blood. You will use special paper, Tallquist paper, and a color guide to estimate hemoglobin concentration.

Calculation of MCV and MCHC
An abnormally low hemoglobin, hematocrit, or RBC count may indicate that a person has anemia. There are many causes and types of anemia and diagnosis of a specific type is aided by relating the above measurements to derive the mean corpuscular volume (MCV) and the mean corpuscular hemoglobin concentration (MCHC). The calculations are as follows:

MCV mean corpuscular volume	$\frac{\text{\% hematocrit} \times 10}{\text{RBC count (millions/mm3)}}$	$\frac{45 \times 10}{5.4} = 83.3\,fl$
MCHC mean corpuscular hemoglobin concentration	$\frac{\text{hemoglobin (g/dl)} \times 100}{\text{\% hematocrit}}$	$\frac{16 \times 100}{46} = 35\%$

* dl = deciliter = 100 ml

Data interpretation: High MCV + normal MCHC suggests **pernicious anemia**, due to folic acid or vitamin B_{12} deficiency. Low MCV + low MCHC suggests **iron–deficiency anemia**, due to inadequate amounts of iron in the diet. Both values will still be normal with **hemorrhagic anemia**, due to acute (wound) or chronic (excessive menstrual flow) blood loss.

RBC measurement	Male	Female
Red cell count (millions/mm^3)	5.4± 0.8	4.8±0.6
Hematocrit (%)	47.0±7.0	42.0±5.0
Hemoglobin (gm/100 ml)	16.0±2.0	14.0±2.0
MCV (fl)	87±5	87±5
MCH (pg)	29±2	29±2
MCHC (%)	34±2	34±2

Table 18.1 – Normal Red Blood Cell Values

Human Physiology Lab 18 Name ______________________________

Blood Lab

1.

	Your Data	Normal? Problem?
RBC count		
hematocrit		
hemoglobin		
MCV		
MCHC		

2. Identify the leukocyte(s) from the following descriptions:

   rarest white blood cell ______________________

   phagocytic white blood cells ______________________

   granules with an affinity for red stain ______________________

   polymorphonuclear with poorly staining granules ______________________

   agranular, round nucleus, relatively little cytoplasm ______________________

3. What is your blood type? ________ What blood type(s) can you receive? ________

Blood Type	List all antigens present on the RBCs.	List all antibodies present in blood plasma.	List all blood types they can receive in a transfusion.	List all recipients they can donate to.
A				
O+				
AB-				

Blood Type	Number of Students	Percent of Students	Percent in USA
O			45
A			42
B			10
AB			3

4. Explain the danger of giving a person with type A blood a transfusion of type B blood. Be specific.

5. Explain how hemolytic disease of a newborn is produced. How may this disease be prevented?

Acknowledgements

The information, tables and figures in this lab manual came from a variety of sources, which are listed below. Since this manual is for class use only, and not to be published, no attempt was made to include quotation marks or references in the body of the text. The tables and figures were given new titles to fit with the laboratory exercise numbers of this manual, but they were taken from the various sources listed below. The primary information source for each lab exercise is given on the following page.

Human Physiology Laboratory Guide, 8th edition, Stuart Ira Fox, McGraw–Hill, 1999.

Experimental and Applied Physiology, 5th edition, Richard G. Pflanzer, Wm. C. Brown Pub., 1995.

Experiments in Physiology, 6th edition, Gerald D. Tharp, MacMillan Pub., 1993.

Human Physiology Laboratory Manual, Sherry Tamone, SSU, 1998.

Human Physiology Laboratory, Ruth Nash, College of Marin, 1992.

Human Physiology, 6th edition, Stuart Ira Fox, McGraw–Hill Pub., 1999.

Human Physiology, from Cells to Systems, 3rd edition, Lauralee Sherwood, Wadsworth Pub., 1997.

Lab Exercises – Primary Source References

Laboratory	Text – primary source
1. Scientific Data	Pflanzer, 1
2. Homeostasis	Fox, 1.3
3. Cells, Tissues	Fox, 1.1, 1.2
4. Enzyme Activity	Fox, 2.4
5. Osmosis	Fox, 2.6A, B, C
6. Nerve Stimulation	Nash
7. CNS Anatomy	Tamone
8. Reflex Arc	Fox, 3.3; Flexicomp
9. Senses	Fox, 3.4, 3.6, 7.9; Tharp, 8
10. Muscle Contraction	Fox, 5.2B; Physiogrip, Carolina Biological Supply Co.
11. Cardiac Function	Fox, 7.2, 5; Cardiocomp
12. Blood Pressure	Fox, 7.6; 7.7
13. Pulmonary Function	Fox, 8.1, 2
14. Renal Function	Fox, 9.1, 3
15. Acid Base Balance	Fox 8.4
16. Digestion	Fox 10.2A, B, C
17. Glucose Tolerance	Pflanzer, 34; Tamone
18. Blood Count, Type	Fox, 6.1A, B; 6.2, 3, 4

Made in the USA
Middletown, DE
14 February 2023